AF410698

Ion-exchange in Glasses and Crystals

Ion-exchange in Glasses and Crystals: from Theory to Applications

Editors

Jesús Liñares Beiras
Giancarlo C. Righini

MDPI • Basel • Beijing • Wuhan • Barcelona • Belgrade • Manchester • Tokyo • Cluj • Tianjin

Editors
Jesús Liñares Beiras
Department of Applied Physics
University of Santiago de
Compostela
Santiago de Compostela
Spain

Giancarlo C. Righini
MiPLab
Nello Carrara Institute of
Applied Physics, CNR
Metropolitan City of Florence
Italy

Editorial Office
MDPI
St. Alban-Anlage 66
4052 Basel, Switzerland

This is a reprint of articles from the Special Issue published online in the open access journal *Applied Sciences* (ISSN 2076-3417) (available at: www.mdpi.com/journal/applsci/special_issues/Ion_exchange_Glasses).

For citation purposes, cite each article independently as indicated on the article page online and as indicated below:

LastName, A.A.; LastName, B.B.; LastName, C.C. Article Title. *Journal Name* **Year**, *Volume Number*, Page Range.

ISBN 978-3-0365-1901-2 (Hbk)
ISBN 978-3-0365-1900-5 (PDF)

Contents

About the Editors

Jesús Liñares Beiras

Jesús Liñares Beiras is a full professor in Optics (since 1998) in the Applied Physics Department, Faculty of Physics, Universidade de Santiago de Compostela (USC). He received BS, MS and PhD (with special distinction) degrees in Physics in 1984, 1985 and 1988, respectively, from the USC. His academic experience is mainly related to subjects such as Classical Optics, Quantum Optics, Integrated Optics, and Interaction Radiation-Matter. Since 2013, he has been the coordinator of the PhD programme Laser, Photonics and Vision (USC). He has authored about 200 international publications in GRIN and fiber optics, integrated optics and quantum optics, and presented several invited papers at international conferences. Since 1998, he has led the research group Quantum Matter and Photonics at the USC. He has been the main researcher responsible for many projects in classical and quantum guided-wave optics and fiber optics. He has also advised many MS and PhD students. He is a member of EOS, SEDO and RSEF.

Giancarlo C. Righini

Giancarlo C. Righini, physicist, is now a senior associate at the Nello Carrara Institute of Applied Physics (IFAC CNR). He worked for more than 40 years at the National Research Council of Italy (CNR), and rose to the position of research director at IFAC. He was the director of the CNR Natl. Group of Quantum Electronics & Plasma Physics (1983–1996) and director of the CNR Natl. Dept. of Materials and Devices (2006–2009). He retired from CNR at the end of 2010, and subsequently became the director of Enrico Fermi Centre, in Rome, Italy (2012–2016). He was the co-founder and then President of SIOF, the Vice President of IUPAP, the Vice President of ICO, the Secretary of EOS, and a member of the Board of Directors of SPIE. He is an honorary chair of the Intl. Workshop PRE—Photoluminescence in Rare Earths. He has authored more than 500 papers, and edited or co-edited 6 books. He is a Fellow of EOS, OSA, SIF, SIOF, and SPIE. Dr. Righini is listed in the top 2% most cited scientists globally within their respective specialty areas .

Preface to "Ion-exchange in Glasses and Crystals: from Theory to Applications"

Ion-exchange (IEx) is a kind of diffusion process where the species in contact are charged particles, such as ions. In such a case, the local electric field generated by the ions' movement imposes some restrictions; thus, only coupled diffusion, namely, the interexchange of two or more ions, is permitted. The IEx process is a natural event, which occurs in soil, minerals, and biological systems. The origins of soil (and IEx) chemistry were in the observations of a few scientists in the early 1800s (among them, the Florentine Giuseppe Gazzeri [1]) and in the detailed studies of the Englishmen H.S. Thompson, in 1850 [2], and J.T. Way, in 1852 [3]. Way, in particular, performed multiple experiments and found that the uptake of K, Ca, and Mg by soils was due to the presence of complex silicates performing an ion-exchange function. The diffusion of Na^+ ions in glass was first studied in 1884 by E. Warburg [4]. The first practical industrial application of the IEx process occurred at the beginning of 1900, and was related to water treatment. It followed up on the studies of two groups of German chemists, namely, F. Harm and A. Rümpler [5] and R. Gans [6,7], who developed synthetic sodium-aluminosilicate exchanger materials. Since then, IEx has become a fundamental process in many applications involving the treatment and purification of water and, more generally, in catalysis, chromatography, and the food and pharmaceutical industries.

A different area where IEx has been utilised for many centuries, even well before the associated chemistry was understood, concerns glassy and ceramic materials. Luster ceramic pottery from Mesopotamia in the 9th century AD and stained glass windows in Gothic churches created in the 14th and 15th centuries in Europe are wonderful examples of ancient artworks produced through an IEx process [8]. Luster decorations, widely used in Medieval and Renaissance pottery in Mediterranean countries, substantially consist of a metal–glass nanocomposite (metal nanoparticles embodied in a silica glassy matrix). This is obtained from the reaction of the pottery glaze with a paste constituting a mixture of ingredients (e.g., copper and/or silver salts and oxides, vinegar, ochre, and clay) painted on its surface. Heating at a temperature of around 500–600 °C in a reducing atmosphere induces an ion exchange process between metals in the paste and alkali ions in the glaze; the Cu^+ and/or Ag^+ ions are then reduced to metal nanoparticles, aggregate, and remain trapped within the first layer of the glaze, producing brilliant reflections of different colors and iridescence.

A similar process has been used since the early 1300s to create a yellow color on clear window glass. The technique, known as silver stain or yellow stain, involves painting silver compounds (e.g., silver oxide or nitrate, sulphate, and chloride) dispersed in a clay medium onto the glass and then firing it in a kiln. During the heating, just above the glass transition temperature, the silver ions are exchanged with the alkali ions present in the glass (mostly Na^+ or K^+). Then, due to the presence of impurities in the glass, silver ions reduce to metallic silver nanoparticles, and the resulting color depends on the size, shape and concentration of the nanoparticles. One of the first documents describing this yellow stain dates back to around 1395–1396, when the glass painter Antonio da Pisa wrote a brief treatise on the various steps of the process of making a painted window [9].

IEx in glass, however, found large-scale industrial applications only at the beginning of 20th century, after the detailed studies of G. Schultze, who was the first, in 1913, to study the process using silver nitrate salt ($AgNO_3$) as an ion source and to outline the induced modifications of the surface of the glass [10]. However, it took almost 50 years more to realise that such changes could be exploited to make the glass more robust: the substitution of small ions such as Li^+ or Na^+ in an alkali-containing glass with larger ions, such as K^+ from a molten KNO_3 bath, induces bi-axial residual compressive stress in the surface layers, which, in turn, strengthens the glassware [11]. K–Na exchange, similarly to Ag–Na exchange, also induces an increase in the refractive index in the diffused layer.

The 1960s and 1970s saw an extraordinary breakthrough in physics and technology, with the invention of laser and the development of low-loss optical fibers, making the design and implementation of ultra-wide-band communication systems feasible. Optical miniaturised components became necessary for the generation, modulation, coupling, switching and detection of optical signals: the article by S.E. Miller, in 1969 [12], was published at the time of pioneering research on planar optical waveguides and the starting point of a new R&D area, namely, integrated optics [13]. Glass has proven to be an excellent material for integrated optical circuits, thanks to its robustness, low propagation losses and low cost; IEx provided a relatively simple and low-cost technology for the fabrication of circuits. After early studies by T. Izawa in Japan [14] and T.G. Giallorenzi in the United States [15], it became clear that ion exchange in glass offered effective solutions to many fabrication problems.

At almost the same time, in 1965, A.A. Ballman at Bell Labs [16] and S.A. Fedulov and collaborators in Russia [17] independently reported the successful growth by the Czochralski technique of large and homogeneous single crystals of lithium niobate and lithium tantalate. These crystals, together with potassium titanyl phosphate (KTP), have acquired great importance in optoelectronics due to their excellent optical properties and their large piezoelectric, electro-optic and nonlinear-optical effects. IEx has proven to be very effective in these crystals as well, for the fabrication of integrated optical elements and devices (e.g., modulators, second-harmonic generators, ring resonators, interferometers, lasers, etc.). The exchangeable ion in these crystals is Li^+, and the most efficient process is an exchange with H^+ (proton exchange), even if Ag/Li exchange in lithium niobate is possible [18].

Due to the relevance of ion-exchange technology for the development of advanced integrated optical components and devices in glasses and crystals, we considered it worthwhile to compile a Special Issue of *Applied Sciences* on this topic. The present volume collects articles published in 2021 [19]. Four papers are reviews and offer a broad overview of the field of IEx in glass, seen from different points of view. Even if ion-exchanged glass waveguides have been already studied for forty years, recently, some advances have been made in the theoretical modelling of the process. Prieto Blanco and Montero Orille [20] present equations that describe the evolution of the cation concentration rewritten in a more rigorous manner. Along with these equations, the boundary conditions for the usual IEx from molten salts, silver and copper films and metallic cathodes were established accordingly. Moreover, the modelling of some IEx processes that have attracted a great deal of attention in recent years, including glass poling, the electro-diffusion of multivalent metals and the formation/dissolution of silver nanoparticles, has been addressed.

The second review article, by Broquin and Honkanen [21], emphasises major breakthroughs in the field of passive and active devices for telecommunication applications. The section dedicated to sensors underlines the evolution of ion-exchange technology, which is developing from quite simple, although extremely performant functions, to more complex integrated optical microsystems.

Berneschi et al. [22] aimed to clearly show how glass and ion-exchange are paired; far from being an obsolete material/technology platform, this still plays a key role in various technological fields with interesting applications and industrial developments that also have repercussions at the level of everyday life. As an example, they underline the role that IEx, together with glass material engineering, can play in two areas: (a) the optimisation of substrates for the development of high-performance surface-enhanced Raman scattering (SERS) devices; and (b) the creation of increasingly high-performing flexible substrates towards the achievement of all-glass flexible photonics as a valid alternative to those developed so far with polymeric materials. The fourth review article, by Righini and Liñares [23], presents an introduction to some fundamental aspects of integrated optical waveguides and devices, such as directional couplers, waveguide gratings, integrated optical amplifiers and lasers, all fabricated by IEx in glass. Then, some promising research activities on IEx glass-integrated photonic devices, and, in particular, quantum devices (quantum circuits), are analysed. According to the increasing interest for passive and/or reconfigurable devices for quantum cryptography or even for specific quantum processing tasks, the implementation of an active integrated quantum state generator device for quantum cryptography and passive devices with an IEx–glass platform is described, such as an integrated quantum projector.

Following these four review papers, the reader may find three original articles. The first, by Montero-Orille et al. [24], proposes a simple polygonal model to describe the phase profile of ion-exchanged gratings. This model enables the design of these gratings, and could also be useful to design more complex diffractive elements. Several ion-exchanged gratings were fabricated to validate the model and to characterise the processes involved in their fabrication; to show the practical utility of this model, the design and fabrication of a grating that removes the zero order and of a three-way splitter are reported, and their performance is analysed. In the study by Nikonorov et al. [25], the influence of small additives on the spectral and optical properties of Na^+–Ag^+ ion-exchanged silicate glass is presented. Polyvalent ions, e.g., cerium and antimony, are shown to reduce silver ions to an atomic state and promote the growth of photoluminescent silver molecular clusters and plasmonic silver nanoparticles. Na^+–Ag^+ ion-exchanged and heat-treated glasses doped with halogen ions, such as chlorine or bromine, exhibit the formation of photo- and thermochromic $AgCl$ or $AgBr$ nanocrystals. The presented results highlight the vital role of small additives to control the properties of silver nanostructures in Na^+–Ag^+ ion-exchanged glasses. Possible applications of Na^+–Ag^+ IEx glass ceramics include, but are not limited to, biochemical sensors based on SERS phenomena, temperature and overheating sensors, white light-emitting diodes, and spectral converters. Finally, Kip et al. [26] report an investigation of the ytterbium diffusion characteristics in lithium niobate. Ytterbium-doped substrates were prepared by the in-diffusion of thin metallic layers coated onto x- and z-cut congruent substrates at different temperatures. The ytterbium profiles were investigated in detail by means of secondary neutral mass spectroscopy, optical microscopy, and optical spectroscopy. Diffusion from an infinite source was used to determine the solubility limit of ytterbium in lithium niobate as a function of temperature. The derived diffusion parameters are of importance for the development of active waveguide devices in ytterbium-doped lithium niobate.

Overall, this volume represents an updated overview of several areas in the field of ion-exchange in glasses and crystals for integrated optics applications. It certainly is not exhaustive, given the high number of papers published on this topic in the forty-year history of integrated optics, but the reader may find sufficient information, covering different topics, such as numerical modelling and the fabrication of ion-exchanged passive waveguides, the design and fabrication of passive and active components and devices, and prospects of applications in optical communications, optical sensing, and quantum photonics. For these reasons, this book may be useful for a broad audience, from MSc and PhD students to early-career researchers and teachers; we hope that it could trigger newcomers'interest and stimulate research to overcome present limitations. A large number of references integrate the physico-chemical descriptions.

All authors are highly acknowledged for contributing to the realisation of this Special Issue; the support from the *Applied Sciences* editorial staff has been greatly appreciated.

References

[1] Gazzeri, G. Compendio d'un trattato elementare di chimica. 3rd ed., Stamperia Piatti, Firenze, Italy, 1828.

[2] Thompson, H. S. On the absorbent power of soils. J. Royal Agricultural Society England 1850, 11, 68-74.

[3] Way, J.T. On the Power of Soils to Absorb Manure. J. Royal Agricultural Society England 1852,13, 123-143.

[4] Warburg, E. Ueber die Electrolyse des festen Glases. Ann. Physik 1884, 21, 622-646.

[5] Harm, F. and Rümpler, V. Internationaler Kongress fur Angewandte Chemie, Berlin. 2.-8. Juni 1903 (Deutscher Verlag, Berlin, 1904).

[6] Gans, R. Zeolites and similar compounds: Their construction and significance for technology and agriculture. Jahrb. Preuss. Geol. Landesanstalt 1905, 26, 179.

[7] Gans, R. Alumino-silicate or artificial zeolite, US. Patent No. 914,405; patented March 9, 1909.

[8] Mazzoldi, P.; Carturan, S.; Quaranta, A.; Sada, C.; Sglavo, V.M. Ion-exchange process: History, evolution and applications. Riv. Nuovo Cim. 2013, 36, 397–460.

[9] Lautie, C.; Sandron, D. Antoine de Pise. L'art du vitrail vers 1400. Éditions du Comité des Travaux Historiques et Scientifiques, Paris, 2008.

[10] Schulze, G. Versuche über die diffusion von silber in glas. Ann. Physik 1913, 345, 335–367.

[11] Kistler, S.S. Stresses in glass produced by nonuniform exchange of monovalent Ions. J. Am. Ceram. Soc. 1962, 45, 59–68.

[12] Miller, S.E. Integrated optics: an introduction. Bell Syst. Tech. J. 1969,

[13] Righini, G.C.; Ferrari, M., Eds., Integrated Optics, The IET, London 2020, 2 volumes.

[14] Izawa, T.; Nakagome, H. Optical waveguide formed by electrically induced migration of ions in glass plates. Appl. Phys. Lett. 1972, 21, 584-586.

[15] Giallorenzi, T.G.; West, E.J.; Kirk, R.; Ginther, R.; Andrews, R.A. Optical Waveguides Formed by Thermal Migration of Ions in Glass. Appl. Opt. 1973,12, 1240-1245.

[16] Ballman AA. Growth of piezoelectric and ferroelectric materials by the Czochralski technique. J. Am. Cer. Soc. 1965, 48,112-113.

[17] Fedulov, S.A.; Shapiro, Z.I.; Ladyzhinskii, P.B. The growth of crystals of LiNbO3, LiTaO3 and NaNbO3 by the Czochralski method. Sov Phys Crystallography 1965,10, 218.

[18] Korkishko, Yu.N. and Fedorov, V.A. Ion exchange in single crystals for integrated optics and optoelectronics. Cambridge Intl. Science Pub., Cambridge, UK, 1999.

[19] https://www.mdpi.com/journal/applsci/special_issues/Ion_exchange_Glasses

[20] Prieto-Blanco, X.; Montero-Orille, C. Theoretical Modelling of Ion Exchange Processes in Glass: Advances and Challenges. Appl. Sci. 2021, 11, 5070. https://doi.org/10.3390/app11115070

[21] Broquin, J.-E. and Honkanen, S. Integrated Photonics on Glass: A Review of the Ion-Exchange Technology Achievements. Appl. Sci. 2021, 11, 4472. https://doi.org/10.3390/app11104472

[22] Berneschi, S.; Righini, G.C.; Pelli, S. Towards a Glass NewWorld: The Role of Ion-Exchange in Modern Technology. Appl. Sci. 2021, 11, 4610. https://doi.org/10.3390/app11104610

[23] Righini, G.C.; Liñares, J. Active and Quantum Integrated Photonic Elements by Ion Exchange in Glass. Appl. Sci. 2021, 11, 5222. https://doi.org/10.3390/app11115222

[24] Montero-Orille, C.; Prieto-Blanco, X.; González-Núñez, H.; Liñares, J. A Polygonal Model to Design and Fabricate Ion-Exchanged Diffraction Gratings. Appl. Sci. 2021, 11, 1500. https://doi.org/10.3390/app11041500

[25] Sgibnev , Y.; Nikonorov, N.; Ignatiev, A. Governing Functionality of Silver Ion-Exchanged Photo-Thermo-Refractive Glass Matrix by Small Additives. Appl. Sci. 2021, 11, 3891. https://doi.org/10.3390/app11093891

[26] Ruter, C.E.; Bruske, D.; Suntsov, S.; Kip, D. Investigation of Ytterbium Incorporation in Lithium Niobate for Active Waveguide Devices. Appl. Sci. 2020, 10, 2189. https://doi.org/10.3390/app10062189

Jesús Liñares Beiras, Giancarlo C. Righini
Editors

Review

Theoretical Modelling of Ion Exchange Processes in Glass: Advances and Challenges

Xesús Prieto-Blanco *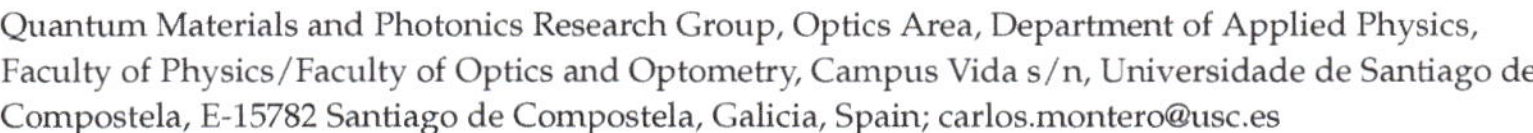 and Carlos Montero-Orille

Quantum Materials and Photonics Research Group, Optics Area, Department of Applied Physics, Faculty of Physics/Faculty of Optics and Optometry, Campus Vida s/n, Universidade de Santiago de Compostela, E-15782 Santiago de Compostela, Galicia, Spain; carlos.montero@usc.es
* Correspondence: xesus.prieto.blanco@usc.es

Abstract: In the last few years, some advances have been made in the theoretical modelling of ion exchange processes in glass. On the one hand, the equations that describe the evolution of the cation concentration were rewritten in a more rigorous manner. This was made into two theoretical frameworks. In the first one, the self-diffusion coefficients were assumed to be constant, whereas, in the second one, a more realistic cation behaviour was considered by taking into account the so-called mixed ion effect. Along with these equations, the boundary conditions for the usual ion exchange processes from molten salts, silver and copper films and metallic cathodes were accordingly established. On the other hand, the modelling of some ion exchange processes that have attracted a great deal of attention in recent years, including glass poling, electro-diffusion of multivalent metals and the formation/dissolution of silver nanoparticles, has been addressed. In such processes, the usual approximations that are made in ion exchange modelling are not always valid. An overview of the progress made and the remaining challenges in the modelling of these unique processes is provided at the end of this review.

Keywords: ion exchange in glass; ion diffusion; glass waveguides; glass strengthening; glass poling; metal nanoparticles

Citation: Prieto-Blanco, X.; Montero-Orille, C. Theoretical Modelling of Ion Exchange Processes in Glass: Advances and Challenges. *Appl. Sci.* **2021**, *11*, 5070. https://doi.org/10.3390/app11115070

Academic Editor: Renato Torre

Received: 20 April 2021
Accepted: 24 May 2021
Published: 30 May 2021

Publisher's Note: MDPI stays neutral with regard to jurisdictional claims in published maps and institutional affiliations.

1. Introduction

Ion exchange in glass has been used for centuries for the purposes of decoration and colouring. Glass lustre on ceramics with metallic nanoparticles from ion exchange has been known from the early Islamic culture during the 10th Century [1]. However, the scientific and industrial application of this technique dates back 60 years ago when potassium ion exchange (IE) was first applied in the chemical surface tempering of glasses [2,3]. Next, with the introduction of the concept of integrated optics in 1969 [4], ion exchange in glass was proposed as a waveguide fabrication process. Just a few years after this proposal, Izawa and Nakagome published the first work on ion exchange waveguides [5]. This kind of waveguide presents several advantages: fibre compatibility, low propagation losses and low cost. Moreover, ion exchange can be combined with other techniques, such as sol–gel, for the fabrication of passive and active (rare-earth-doped) integrated optical devices [6]. Currently, the main applications of ion exchange are in glass strengthening [7–11] and in the fabrication of photonic components for both guided-wave [12–18] and bulk optics [19–21].

In an IE process, cations (mostly Na^+) close to the glass surface are replaced with other monovalent cations such as K^+, Li^+, Rb^+, Cs^+, Tl^+ or Ag^+ [22]. Molten salts of such cations are common sources of dopants, although a metallic film deposited on the glass surface can be also a source of cations. The exchange takes place by a purely thermal diffusion process or it can be assisted by an electric field. The new cations can change the electrical permittivity, the stress and even the absorption of the glass, these being changes proportional to the dopant cation concentration. By selective masking of the glass surface, ion exchange can be locally prevented or allowed, giving rise to custom-made elements for specific purposes. The theoretical modelling of the IE processes ran parallel to the development of the technologies. This modelling was fundamental to the design and fabrication of most elements. For instance, in the field of integrated optics, some subjects as fibre compatibility or coupling losses depend strongly on the permittivity distribution and, hence, on the cation concentration. Therefore, the prediction of the cation concentration is of great importance to design devices based on IE technologies.

Here, we present a review of the theoretical modelling of IE processes in glass made up to now, as well as the last advances in this matter that have been reported in recent years. Ion exchange within a glass network is governed by diffusion and drift processes as a response to a concentration gradient and an electric field, respectively [23,24]. This gives rise to a concentration profile of the exchanging ions, which depends on the processing conditions of the substrate: temperature, exchange time, applied electric field, etc. The Nernst–Planck drift-diffusion equation describes this process. It establishes the proportionality among the flux density of each ion species and both the electric field and the concentration gradient. On the other hand, Poisson's equation and continuity equations for each ion must be fulfilled. This gives rise, in general, to three second-order coupled partial differential equations whose solution provides the evolution of both the cation concentrations and the electric potential. In the derivation of these equations, the charge neutrality approximation is usually assumed, which allows for some important simplifications without a relevant loss of accuracy in the most common cases [25,26]. All of these subjects are addressed in Section 2, where we present the basic model of ion exchange. This model assumes that the self-diffusion coefficients of the exchanged cations are constant for a given temperature. However, experimental measurements [27] showed that these coefficients depend on the cation concentrations. Therefore, this basic model was generalized to non-ideal cation behaviour and concentration-dependent self-diffusion coefficients by considering the so-called mixed ion effect [28,29]. Later, both models were rigorously generalised, and in the derivation of the equations, Faraday's law was considered instead of Ohm's law [30]. This leads to a non-standard Laplace equation for an effective electric potential. Accordingly, the boundary conditions for the most common IE processes were established. These conditions, together with the aforementioned partial differential equations, complete the theoretical modelling of the IE problem. On the other hand, some IE processes have attracted a great deal of attention in the last few years due to their remarkable applications. Among these, we must highlight: glass poling, electro-diffusion of multivalent metals and the formation/dissolution of silver nanoparticles. Their modelling has only been partially done so far [31–33], because the usual theoretical assumptions (mainly charge neutrality approximation and ideal cation behaviour) are not always valid in such processes. In Section 4, we give an overview of the progress made and the challenges still to be faced on this matter.

2. Basic Model

The simplest problem of ion exchange arise when two species of monovalent cations (A and B) exchange with each other in the same glass region at a given temperature T. Let us consider an infinite one-dimensional medium (x being the spatial coordinate) with a homogeneous concentration C_0 of fixed anions (typically -Si-O$^-$ radicals) and non-homogeneous and variable cation concentrations $C_A(x,t)$ and $C_B(x,t)$, t being the time. Moreover, we considered that each cation is initially near its anion, that is they are paired, so $C_A(x,0) + C_B(x,0) = C_0$. Consequently, space charge density is initially cancelled:

$$C_0 = C_A + C_B ,\qquad(1)$$

that is local charge neutrality is met. Our goal was to obtain the evolution of cation concentrations given these initial conditions.

2.1. Nernst–Planck and Poisson Equations

The initial non-homogeneity of cation concentrations, $C_A(x, t)$ and $C_B(x, t)$, and their random motion make them diffuse along the glass until their concentrations are homogeneous. This diffusion process is described by Fick's law:

$$J_i^{dif} = -D_i \nabla C_i \qquad i = A, B \, ; \tag{2}$$

where J_i is the flux density and D_i the diffusion coefficient of each cation. However, the two interdiffusing cations have usually different diffusion coefficients and, therefore, different mobilities, which produce charge imbalances. These imbalances generate a strong internal electric field (E), which tends to balance the charges and restore charge neutrality (Equation (1)). Therefore, Fick's equation is no longer valid, and a drift term must be added on its right-hand side to take into account the effect of this electric field on the cation motion. This leads to the Nernst–Planck equation [34]:

$$J_i = -D_i \nabla C_i + D_i C_i \frac{eE}{kT} \qquad i = A, B \, ; \tag{3}$$

where e is the proton charge, T the absolute temperature and k Boltzmann's constant. Note that some authors included, in the drift term of this equation, the Haven ratio. However, as we will see below, this parameter should not be incorporated into the model as a general rule. The occurrence of the above-mentioned electric field can be seen from a quantitative point of view by calculating the total flux density:

$$J_0 = J_A + J_B = -D_A \nabla C_A - D_B \nabla C_B + (D_A C_A + D_B C_B) \frac{eE}{kT} \tag{4}$$

which depends on the electric field through mobility u:

$$u = \frac{e}{kTC_0}(D_A C_A + D_B C_B). \tag{5}$$

However, J_0 cancels in the current problem, that is,

$$J_0 = 0, \tag{6}$$

because there is no external field applied that generates a net current. This means that the imbalance of the diffusion of the two cations ($D_A \nabla C_A + D_B \nabla C_B$) is compensated by the internal field through the drift term uC_0E. Note that Equation (4) is a generalization of Ohm's law. On the other hand, as long as there is no creation or destruction of cations from/to a metallic state, the continuity equation must be fulfilled:

$$\frac{\partial C_i}{\partial t} + \nabla J_i = 0 \qquad i = A, B. \tag{7}$$

Now, by combining this equation and Equation (3) for, for instance, $i=A$ and substituting, in the resulting equation, the electric field from Equations (4) and (6), we obtained the differential equation that gives the concentration evolution [23] of cation A:

$$\frac{\partial c_A}{\partial t} = \nabla(\bar{D}(c_A)\nabla c_A), \tag{8}$$

where the charge neutrality (Equation (1)) was applied and a normalized concentration ($c_A = C_A / C_0$) was used, and we defined the following interdiffusion coefficient:

$$\bar{D}(c_A) = \frac{D_A}{1 - \alpha c_A}, \tag{9}$$

where $\alpha = 1 - D_A/D_B$. The dependence of u and $\bar{D}$ on c_K predicted by this basic model, for K^+/Na^+ IE, can be seen in Figure 1a.

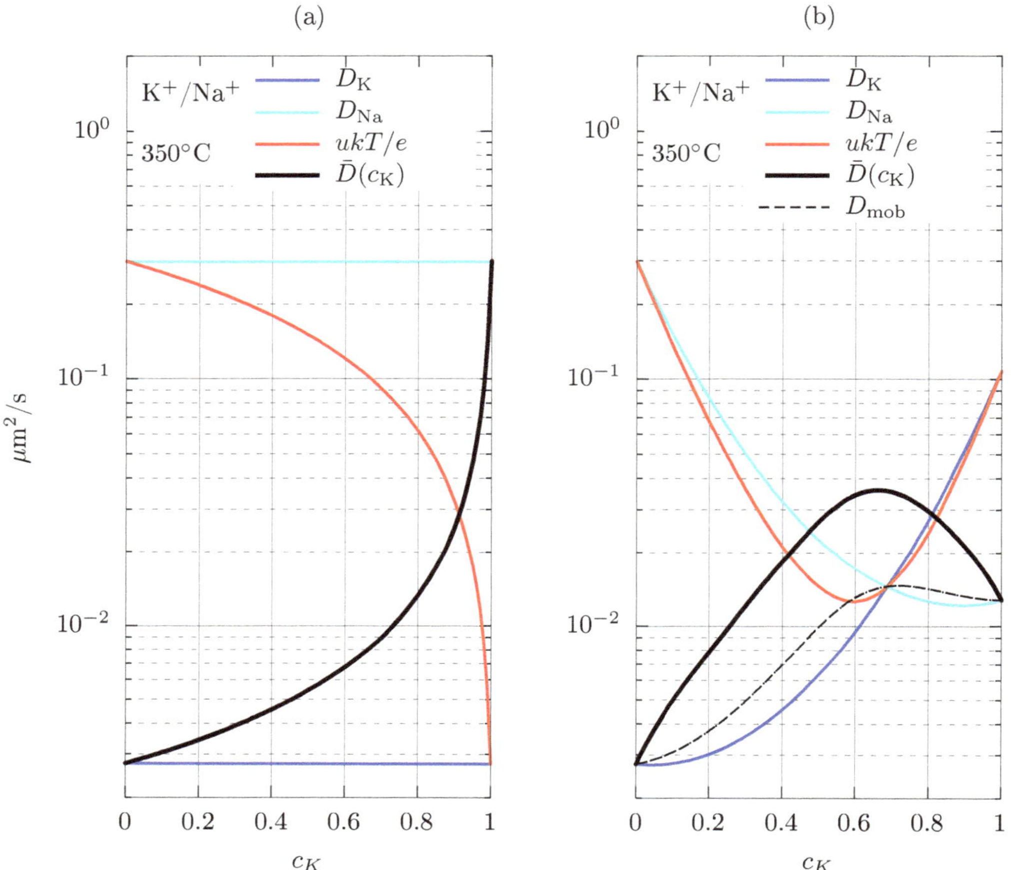

Figure 1. Self-diffusion and interdiffusion coefficients, as well as mobility, calculated from experimental data obtained by radiative tracers [27], as a function of the normalized concentration for K^+/Na^+ IE. The Haven ratio was ignored. (**a**) Basic model—Equations (5) and (9)—which assumes that the self-diffusion coefficients remain constant with the cation mole fraction. (**b**) The same functions taking into account the MIE. Quadratic polynomials in c_K were fitted to the logarithms of the experimental self-diffusion coefficients and then used to calculate the rest of the functions through the definitions (35) and (34). Note the difference between $\bar{D}$ and D_{mob}, which are the expected interdiffusion coefficients when the interaction among cations or the ideal mixture is assumed, respectively.

A more complex problem is the IE assisted by an external electric field. In such a case, Equation (6) is not met. Therefore, additional equations are necessary to calculate the total flux density (J_0) or, alternatively, E, which now includes the external field. As for E, it must fulfil Poisson's equation and Faraday's law of induction:

$$\nabla(\epsilon E) = e(C_A + C_B - C_0) \tag{10}$$

$$\nabla \times E + \frac{\partial B}{\partial t} = 0 \,, \tag{11}$$

where ϵ is the glass electrical permittivity and $\mathbf{B}$ is the magnetic field, which will be assumed as time independent since the total current changes very slowly. On the other hand, we also assumed the charge neutrality approximation (Equation (1)). This cannot be done in general, due to the existence of the aforementioned external field; however, in most cases, this is a very good approximation (see the next subsection). Doing this, the addition of continuity Equation (7) leads to:

$$\nabla J_0 = 0. \tag{12}$$

Now, by finding the electric field in Equation (4):

$$\frac{e\mathbf{E}}{kT} = \frac{J_0}{D_A C_A + D_B C_B} + \frac{D_A \nabla C_A + D_B \nabla C_B}{D_A C_A + D_B C_B}, \tag{13}$$

taking into account Equation (12) and doing the same steps as before, the following electro-diffusion equation is obtained:

$$\frac{\partial c_A}{\partial t} + \frac{J_0}{C_0} \frac{1-\alpha}{(1-\alpha c_A)^2} \nabla c_A = \nabla(\bar{D}(c_A)\nabla c_A), \tag{14}$$

which is an extension of Equation (8). On the other hand, Equation (13) can be inserted into (11), leading to:

$$\nabla \times \left(\frac{J_0}{D_B C_0 (1 - \alpha c_A)} \right) = \mathbf{0}, \tag{15}$$

where we used Equation (9). This equation, Equation (12), and the boundary conditions, which will be presented later, determinate the flux density J_0. From this flux density, Equation (14) will provide the evolution of the concentration of cations. Finally, once J_0 and c_A are known, the charge neutrality approximation can be checked. This will be analysed in the following subsection.

Alternatively, the last four equations can be expressed in terms of a scalar function. Indeed, an irrotational vector field is the gradient of a potential function ϕ, so:

$$\frac{J_0}{D_B C_0 (1 - \alpha c_A)} = -\frac{e}{kT} \nabla \phi, \tag{16}$$

where the factor $-e/(kT)$ was included in order for ϕ to have the same units as the electric potential. From this equation and Equation (12), we obtained a non-standard Laplace equation:

$$\nabla((1 - \alpha c_A)\nabla \phi) = 0, \tag{17}$$

which is more convenient for resolution purposes than vector Equations (12) and (15). Note that this effective potential ϕ is not the electric potential V, whose minus gradient is the electric field $\mathbf{E}$ given by Equation (13). However, a relationship between them can be obtained [30] by inserting Equation (16) into Equation (13), that is:

$$\phi = V + \frac{kT}{e} \ln(1 - \alpha c_A). \tag{18}$$

Under typical IE conditions, the difference between V and ϕ is no greater than a few tenths of a volt, which is negligible compared to the usual voltages (20–100 V) used in field-assisted IE. Despite this, this difference must not be ignored in the modelling, as significant errors in the calculation of the electric field could be made. Indeed, although both potentials are similar, their gradient is not always.

On the other hand, it is worth mentioning that Equations (12) and (15) are trivially solved in the one-dimensional case, that is J_0 is constant. This constant will be established from the experimental setup. If a constant current source is used, this value is set directly.

Otherwise, when the external field is generated by a constant voltage source, the value of J_0 can be calculated from the voltage applied to the sample and Equations (14) and (16)–(18).

2.2. Charge Neutrality Approximation in Field-Assisted Ion Exchange

In the previous subsection, the charge neutrality was assumed in the field-assisted IE problem, as well as in the thermal-only case. However, this charge neutrality is not always fulfilled, especially when strong external electric fields are used and/or very high concentration gradients exist. In fact, some authors have modelled the silver concentration in channel waveguides by considering explicitly the space charge distribution [35], albeit at the cost of including other assumptions.

An estimation of the validity of the charge neutrality approximation can be made from Equation (13), which can be expressed as:

$$\frac{eE}{kT} = \nabla \ln(C_A + C_B) + \frac{\frac{J_0}{(C_A+C_B)D_B} - \nabla c_A}{1 - \alpha c_A} \simeq \frac{\frac{J_0}{C_0 D_B} - \nabla c_A}{1 - \alpha c_A}, \tag{19}$$

where we used the charge neutrality approximation and the definition of α. Now, we took divergences in this equation in order to compare it with Equation (10) and estimate the error made by this approximation. This is an iterative procedure. First, we used this approximated expression for the electric field. Next, we substituted it into Poisson's Equation (10) to obtain a more accurate value for $C_A + C_B$, which was introduced in the previous equation, and so on. However, a unique iteration will be enough to obtain an order of magnitude of the charge density [36]. Therefore, if we use Equation (12) and assume that ϵ does not depend on C_A and C_B, we obtained:

$$\frac{e^2}{\epsilon kT}(C_A + C_B) - \frac{e^2 C_0}{\epsilon kT} \simeq \frac{\alpha J_0 \nabla c_A}{(1 - \alpha c_A)^2 C_0 D_B} - \left(\frac{\alpha \nabla c_A}{1 - \alpha c_A}\right)^2 - \frac{\nabla^2 c_A}{1 - \alpha c_A}. \tag{20}$$

Now, from this equation, we can obtain some conditions for the validity of the charge neutrality approximation by comparing each term on the right-hand side of this equation with the second term on the left-hand side. Therefore, for the second term on the right-hand side, this comparison provides:

$$\frac{\alpha \nabla c_A}{1 - \alpha c_A} \ll \sqrt{\frac{e^2 C_0}{\epsilon_0 \epsilon_r kT}} \simeq 1.1 \times 10^{10} \text{ m}^{-1}, \tag{21}$$

for a BK7 glass with a density of 2.4 g/cm^3 and 8.4% by weight of Na$_2$O, which gives $C_0 \simeq 3.9 \times 10^{27}$ m^{-3}; likewise, typical values for the temperature, T = 400 °C, and for the relative permittivity, ϵ_r = 10, were considered. As for the ion exchange, we chose for this assessment a Na$^+$/K$^+$ IE in a soda-lime glass with a diffusion coefficient ratio $D_K/D_{Na} \simeq 2.5 \times 10^{-2}$ [37], which leads to α = 0.975. Therefore, ∇c_A must fulfil the following condition:

$$\nabla c_A \ll 0.28 \text{ nm}^{-1} \tag{22}$$

In the worst case, c_A = 1. This condition means a change, in the normalized concentration, from one to zero-point-nine at a distance much greater than 0.36 nm, which in practice is met in almost any IE process. Indeed, most IE processes use glasses with a higher Na$_2$O content or their α values are less than the one of Na$^+$/K$^+$ IE. Therefore, this condition is even less restrictive in those cases.

On the other hand, the third term on the right-hand side of Equation (20) is small under the same conditions as the second one. Finally, the first one is small if the second one is and, furthermore, if:

$$\frac{eJ_0}{1 - \alpha c_A} \ll \frac{e^2 C_0^{3/2} D_B}{\sqrt{\epsilon_0 \epsilon_r kT}} \sim 412 \text{ A/cm}^2, \tag{23}$$

where we took into account [37] that $D_B \simeq 6 \times 10^{-13}$ m^2/s, for T = 400 °C. That is, in the worst case ($c_A = 1$), the current density must fulfil:

$$eJ_0 \ll 10 \text{ A/cm}^2. \tag{24}$$

This condition is ensured in practice because the highest current densities that have been used in field-assisted IE processes until now are much lower than this value. In fact, Joule heating should be taken into account for currents higher than a few mA/cm^2, due to the strong dependence of diffusion coefficients on the temperature [38,39]. Therefore, the charge neutrality approximation is valid in most IE process, even for Na$^+$/Rb$^+$ and Na$^+$/Cs$^+$ ion exchanges, which give rise to very steep concentration profiles. In these kinds of profiles, although ∇c_A may be very high, the denominator of Equations (21) and (23) is not close to zero, even for molar fractions close to one. Indeed, we assumed that $\alpha = 1 - D_A/D_B$ is constant, but actually, the diffusion coefficients depend on the molar fractions, so that if a cation has a low concentration, it also has a lower diffusion coefficient. In other words, if $c_A \simeq 1$, then $\alpha < 0$, and $1 - \alpha c_A$ is never close to zero. The basic model that we present in this section cannot explain these dependencies on the concentration of the diffusion coefficients, and therefore, this cannot be used to describe the whole range of molar fractions.

2.3. Boundary Conditions

When an external medium is in contact with the glass surface, each type of cation present in the glass and/or in the medium can either cross or not cross the glass/medium interface. This depends on the nature of such a medium. If a cation crosses the interface, a thermodynamic equilibrium is assumed between the crossing cations at both sides. Otherwise, the normal component of its flux density cancels. Therefore, the different boundary conditions for the Equations (14), (16) and (17) are obtained for a mask/air, a silver or copper film and a molten salt mixture, when none, one or both cations, respectively, cross the interface. Moreover, when the external medium is a conductor (metallic film or fused salt), its electric potential can also be chosen and directly affects the boundary condition of the potential ϕ through Equation (18).

In Tables 1 and 2, we summarize the boundary conditions for IE processes from molten salt mixtures and films, as well as under a mask or in air. The equations that govern these processes are also shown. Note that boundary conditions for other common IE processes as annealing or secondary ion exchanges are included in these cases. Indeed, boundary conditions for annealing processes are the same as for "mask/air", and the ones for secondary ion exchanges are included in the "salt" case, just using a different constant "C", which depends on the dopant concentration in the salt mixture and temperature through an equilibrium equation [40–43]. Moreover, the final concentration of the first process is the initial condition for the second one.

We show, in these tables, the two alternative forms presented in Section 2.1. In Table 1, the problem is formalized in terms of the normalized concentration of dopant cations c_A and the total flux density J_0. This is a more straightforward form, which makes it easier to understand the physical problem of ion exchange. However, for resolution purposes, the form given in Table 2 is simpler because it is written in terms of c_A and a scalar function ϕ, which can be seen as an effective electric potential [30].

The basic model of this section implicitly assumes an ideal cation behaviour in the glass. Therefore, for the sake of consistency, the same assumption was made in the electrochemical potentials to obtain the theoretical equations for the boundary conditions of Tables 1 and 2. However, the corresponding condition for c_A in a glass–salt interface, as a function of the salt composition, must contain non-ideal terms on the glass side, as we will show in the next section. Owing to the molten salt being homogeneous, $c_A|_S$ is constant along the surface. Hence, a mixed theory can be a pragmatic approach; similarly, experimental values for $c_A|_S$ can be used to feed the numerical algorithms that model the ion exchange. On the other hand, the boundary conditions for ϕ are sometimes irrelevant (thermal diffusion) or they can be approximated to those of V (diffusion assisted with strong fields). As a result, they were often ignored in models and experiments in the context of waveguide fabrication.

Table 1. Complete formulation of the IE problem (equations and boundary conditions), for the most common IE processes, in terms of the normalized concentration of dopant cations c_A and the total flux density J_0; $\hat{e}_S$ is a unit vector perpendicular to the sample surface, and C is a constant, which mainly depends on the dopant concentration in the salt.

	c_A	J_0
Equations	$\dfrac{\partial c_A}{\partial t} + \dfrac{J_0}{C_0}\dfrac{1-\alpha}{(1-\alpha c_A)^2}\nabla c_A = \nabla(\bar{D}(c_A)\nabla c_A)$	$\begin{cases} \nabla J_0 = 0 \\ \nabla \times \left(\dfrac{J_0}{D_B C_0(1-\alpha c_A)} \right) = 0 \end{cases}$
Salt	$c_A\vert_S = C$	$\begin{cases} J_0 \times \hat{e}_S = 0 \\ \int_{I_i} J_0 dS = I_i \end{cases}$
Film	$\nabla c_A\vert_S \cdot \hat{e}_S = \dfrac{(c_A\vert_S - 1)\,J_0\vert_S \cdot \hat{e}_S}{D_A C_0}$	$\begin{cases} J_0 \times \hat{e}_S = C_0 D_B \dfrac{\nabla c_A \times \hat{e}_S}{c_A} \\ \int_{I_i} J_0 dS = I_i \end{cases}$
Mask/Air	$\nabla c_A\vert_S \cdot \hat{e}_S = 0$	$J_0 \cdot \hat{e}_S = 0$
with	$\sum_i I_i = 0$	

Table 2. Complete formulation of the IE problem (equations and boundary conditions), for the most common IE processes, in terms of the normalized concentration of dopant cations c_A and an effective electric potential ϕ; $\hat{e}_S$ is a unit vector perpendicular to the sample surface, and C and F are constants.

	c_A	ϕ
Equations	$\dfrac{\partial c_A}{\partial t} + \dfrac{J_0}{C_0}\dfrac{1-\alpha}{(1-\alpha c_A)^2}\nabla c_A = \nabla(\bar{D}(c_A)\nabla c_A)$	$\nabla((1-\alpha c_A)\nabla\phi) = 0$
Salt	$c_A\vert_S = C$	$\phi\vert_S = F$
Film	$\nabla c_A\vert_S \cdot \hat{e}_S = \dfrac{(c_A\vert_S - 1)\,J_0\vert_S \cdot \hat{e}_S}{D_A C_0}$	$\phi\vert_S = F + \dfrac{kT}{e}\ln\left(\dfrac{1}{c_A\vert_S} - \alpha\right)$
Mask/Air	$\nabla c_A\vert_S \cdot \hat{e}_S = 0$	$\nabla\phi\vert_S \cdot \hat{e}_S = 0$
with	$\dfrac{J_0}{D_B C_0} = -\dfrac{e}{kT}(1-\alpha c_A)\nabla\phi$	

2.4. Some Particular Solutions

Two noticeable one-dimensional solutions are obtained for both specific thermal and field-assisted IE processes. In the simplest thermal IE, a glass sheet is immersed in a molten salt, which contains foreign cations. The boundary conditions for c_A and ϕ are constant over time and along the glass surface, then $J_0 = 0$. The concentration profile of the dopants scales in depth proportionally to the square root of the diffusion time, but it retains its shape, that is $c(x,t) = c_A\vert_S f(x/\sqrt{4D_A t})$, being the origin of the coordinates at the glass surface [44]. The shape of f depends on $c_A\vert_S$ and α. If either of them is small, the diffusion Equation (14) becomes linear and f tends to a complementary error function (erfc), which has an inflection point at the glass surface and decreases monotonically to zero inside the glass. Otherwise, the diffusion is faster near the surface, where the dopant concentration is high. This causes f to present a bump near the surface, being the inflection point displaced into the glass. f is often approximated by a Gaussian function, although other functions have been proposed [45–47].

When a voltage is applied between both surfaces of the glass sheet, a current normal to these surfaces appears. If J_0 is kept constant, a stationary solution of Equation (14) exists when the slow cations invade a region containing a higher concentration of fast cations [24]. In that case, the mole fraction profile of the slow cations is a step-like function that moves at a constant speed while maintaining its shape. Therefore, two regions with different concentrations are formed with an abrupt transition between them. The higher the step is, the sharper the front and the greater its velocity because the mobility of slow cations increases with its concentration. The stability results from a competition between diffusion and the non-linearity of the drift term. The former widens the profile, whereas the latter sharpens it because the rear region of the profile moves faster. On the other hand, if V is kept constant, J_0 decreases slowly over time, since the resistivity of the sample increases as the doped region becomes thicker. This effect may already be appreciable for doped regions a few microns deep, because the mobility ratio can reach several orders of magnitude [38,48]. In [38], the applied voltage V had to be corrected by 0.93 V to accurately describe the experimental depths because these models did not include the potential ϕ. Note that a combination of the boundary conditions of ϕ at both glass sides, as well as the difference between ϕ and V given by Equation (18) explain such a potential. Finally, it is worth noting that a stable profile is only formed if the slow cation chases the fast one. Otherwise, the profile extends indefinitely.

3. The Mixed Ion Effect

Constant diffusion coefficients were assumed in Section 2. However, they depend on both the temperature T and the molar fraction c_A. In monoalkaline glasses, the temperature dependence follows the Arrhenius law:

$$D = D_0 \exp\left(-\frac{Q}{kT}\right). \tag{25}$$

This is due to each cation needing to surmount a potential barrier of height equal to Q to move to another potential well. The diffusion coefficient is proportional to the number of cations that overcome this energy in a given instant. This number follows a Maxwell–Boltzmann statistic, which explains the above exponential law. According to the Stuart–Anderson model [49], the potential barrier has both an electrostatic component, corresponding to the energy necessary to separate the cation from its anion, and a mechanical component that describes the glass network distortion necessary for the cation to break through another potential well. Therefore, when two species of cations are present in the glass, it is natural that each one has its own diffusion coefficient. However, surprisingly, the D_0 and Q values of each species (D_{0A}, Q_A and D_{0B}, Q_B) largely depend on the cation mole fraction, in such a way that each diffusion coefficient is reduced by up to several orders of magnitude as its respective mole fraction is approaching zero [27,50,51]. This reduction is mainly due to an increase of the activation energy of minority cations; furthermore, it is stronger than the difference between the diffusion coefficients of both species in their respective monoalkaline glasses ($D_{AA} \equiv D_A|_{c_A=1}$ and $D_{BB} \equiv D_B|_{c_A=0}$). Consequently, D_A and D_B are equal for some intermediate mole fractions. We will show that this leads to a maximum value of the interdiffusion coefficient and a minimum value of the direct current (DC) conductivity for this intermediate mole fraction or its neighbourhood. Therefore, the DC conductivity shows an excess of activation energy with respect to the monoalkali glasses (this excess disappears with frequency in AC conductivity). This phenomenon, among others, is included in the so-called double-alkali effect, mixed alkali effect or, later, mixed mobile ion effect or mixed ion effect (MIE), since silver, thallium or copper cations behave similarly to alkali ones in glass [37,52–54].

3.1. Brief Review of Theories

The origin of the MIE has been debated for a long time, but no theory has been universally accepted yet. Some theories focus on an interaction among neighbouring cations in such a way that mixed pairs are assumed to be more energetically stable than pairs of cations of the same species [15,55–59]. This assumption is supported by several experimental achievements. Namely, the interdiffusion coefficient presents a thermodynamic term for alkali IE [60,61] and for Ag^+/Na^+ exchange [62]; we will show that this term is missing if cations behave as an ideal gas. Furthermore, the surface mole fraction of cations in ion exchanged glasses from molten salts (the salt boundary condition) must be explained on the assumption that a non-ideal cation behaviour both for double-alkali exchange [40,41] and for Ag^+/Na^+ exchange [42,43]. On the other hand, mixing enthalpy experiments do not have a clear interpretation. A negative mixing enthalpy (net attraction among dissimilar cations) was found [63–65], which correlates linearly with the excess of activation energy for DC conduction. However, the former is about 20 times weaker than the latter. Besides, mixed pairs of cations would be expected to be much more likely than pairs of the same species in the presence of interaction, but nuclear magnetic resonance experiments [66–68], as well as neutron and X-ray diffraction [69] show that they are rather randomly distributed. Consequently, other authors attributed the MIE to relaxation processes in the glass structure [70–73]. In particular, each cation is assumed to modify its neighbourhood after a relaxation time to achieve an energetically favourable site. Therefore, in a monoalkaline glass, all the sites are of the same type. Once a foreign cation enters this glass and modifies a site, it is difficult for it to diffuse because all accessible sites are the wrong type. Even if the cation gains access to one of them, most likely, it will return before the new site relaxes. Moreover, diffusion is assumed to occur through conduction pathways, so a foreign cation can block several indigenous ones. This explanation is compatible with the ideal behaviour of cations, that is with a random distribution.

The theory that we assumed is relevant because, depending on it, the resulting diffusion equation is slightly different, as we will show below.

3.2. The Cation Flux Density

Let us consider the electrochemical potential ($\tilde{\mu}_i$) of each cation species that has a chemical term depending on the thermodynamic activity (a_i) of that species and an electric term proportional to the cation charge (e) and the electric potential (V):

$$\tilde{\mu}_i = \mu_i^0 + kT \ln a_i + eV \qquad i = A, B, \tag{26}$$

The flux density of each cation species is proportional to both the gradient of its electrochemical potential and its concentration. This leads to the Nernst–Planck equation:

$$J_i = -D_i C_i \nabla \tilde{\mu}_i = -g_i D_i \nabla C_i + D_i C_i \frac{eE}{kT} \qquad i = A, B, \tag{27}$$

where the g_i's are the thermodynamic factors:

$$g_i \equiv \frac{\partial \ln a_i}{\partial \ln c_i} \qquad i = A, B.$$

As mentioned above, some authors include the Haven ratio in the drift term of Equation (27). However, the Haven ratio should only be used to obtain D_i from the experimental values of the diffusion coefficient of radioactive tracers: D_i^* [74]. The difference between them arises, for example, if the diffusion mechanism is the indirect interstitial one. In this case, a cation located in one interstice replaces a nearby regular site cation, which jumps to another interstitial site. Consequently, the total mass (or charge) displacement described by D_i is different from that of a single cation, which is measured by the tracer. Moreover, the tracer is also affected by the thermodynamic factor, then a comparison between the mobility of a species and the corresponding tracer diffusion coefficient can result in an apparent Haven ratio [61].

If the MIE is fully due to the relaxations of the glass structure, the cation behaviour being ideal, the activity will be equal to the mole fraction, so $g_i = 1$. On the contrary, if the cation interaction is the only thing responsible for the MIE, we will obtain $D_i = D_{ii}\gamma_i$, γ_i being the thermodynamic activity coefficient ($a_i = \gamma_i c_i$). Let us deduce that result. We divided all cations A into two sets, the ones that are hopping from one site to another in a given instant ($A^\uparrow$) and the rest of them, which are fixed ($A^\downarrow$). Although the former are much scarcer, both sets are in thermal equilibrium in any given small region in the glass; therefore:

$$\tilde{\mu}_A = \tilde{\mu}_{A\downarrow} = \tilde{\mu}_{A\uparrow} \tag{28}$$

Now, we made two assumptions. First, we supposed that mobile cations behave ideally with respect to each other due their low concentration ($C_{A\uparrow} \ll C_{A\downarrow}$):

$$\tilde{\mu}_{A\uparrow} = \mu^0_{A\uparrow} + kT \ln \frac{C_{A\uparrow}}{C_0} + eV \qquad i = A, B. \tag{29}$$

Note that this could fail in the case of cooperative movement, that is, when two or more nearby cations change their site simultaneously. Second, we assumed that the reference potential of mobile cations ($\mu^0_{A\uparrow}$) is independent of c_A; therefore, $\nabla \mu^0_{A\uparrow} = 0$. In addition to using Equation (27), the cation flux density can also be calculated from mobile cations, being proportional to both the gradient of their electrochemical potential and their concentration:

$$J_A = J_{A\uparrow} = -\frac{D_{A\uparrow}}{kT} C_{A\uparrow} \nabla \tilde{\mu}_{A\uparrow} = -D_{A\uparrow} \nabla C_{A\uparrow} + \frac{eD_{A\uparrow}}{kT} C_{A\uparrow} E, \tag{30}$$

where $D_{A\uparrow}$ is a temperature-dependent multiplicative coefficient and the factor $1/kT$ was introduced in order for $D_{A\uparrow}$ to have units of a diffusion coefficient. By combining Equations (26), (28) and (29), we can find $C_{A\uparrow}$ as:

$$C_{A\uparrow} = \exp\left(-\frac{\mu^0_{A\uparrow} - \mu^0_A}{kT}\right) \gamma_A C_A,$$

and replace it in Equation (30). The resulting flux density is:

$$J_A = D_{A\uparrow} \exp\left(-\frac{\mu^0_{A\uparrow} - \mu^0_A}{kT}\right) \left\{ -\left(\gamma_A + C_A \frac{\partial \gamma_A}{\partial C_A}\right) \nabla C_{A\uparrow} + \frac{e}{kT} \gamma_A C_A E \right\}$$

which agrees with Equation (27) by identifying:

$$D_A = D_{A\uparrow} \exp\left(-\frac{\mu^0_{A\uparrow} - \mu^0_A}{kT}\right) \gamma_A = D_{AA} \gamma_A.$$

Obviously, an identical derivation can be done for B cations to obtain $D_B = D_{BB} \gamma_B$.

Surprisingly, it was not necessary to make any assumptions about the particular dependence of γ_i on the mole fraction. In short, under these cation interaction assumptions, the dependence of the diffusion coefficient of each species on the mole fraction was directly related to its thermodynamic activity coefficient. For the equations to remain valid, regardless of the MIE explanation, we will continue the derivation from Equation (27), without any assumptions on the thermodynamic term g_i, which can be done in the last step.

3.3. Generalized Equations and Boundary Conditions

By following the same procedure as in the previous section, we replaced E with J_0 in the expression of J_A, and then, we applied the continuity condition to obtain:

$$\frac{\partial c_A}{\partial t} + \frac{\partial}{\partial c_A}\left[\frac{D_A c_A}{D_A c_A + D_B c_B}\right] \frac{J_0}{C_0} \nabla c_A = \nabla \left(\frac{(g_B c_A + g_A c_B) D_A D_B}{D_A c_A + D_B c_B} \nabla c_A\right), \tag{31}$$

where the whole flux density J_0 can be obtained from the scalar potential ϕ as:

$$J_0 = -(D_A c_A + D_B c_B)\frac{eC_0}{kT}\nabla\phi, \tag{32}$$

and ϕ in turn satisfies the following non-standard Laplace equation:

$$\nabla[(D_A c_A + D_B c_B)\nabla\phi] = 0. \tag{33}$$

In view of Equation (31), we can redefine the interdiffusion coefficient as:

$$\bar{D}(c_A) \equiv \frac{(g_B c_A + g_A c_B)D_A D_B}{D_A c_A + D_B c_B} \equiv (g_B c_A + g_A c_B)D_{\text{mob}}. \tag{34}$$

It can be split into a mobility term D_{mob} and a thermodynamic term $(g_B c_A + g_A c_B)$, which is not in the definition (9) of the basic model. If cations behave ideally, the latter becomes equal to one, and the interdiffusion coefficient is reduced to the mobility term. If cation interactions are relevant, the thermodynamic term enhances the maximum of $\bar{D}(c_A)$ for intermediate values of c_A, as can be seen in Figure 1b. Similarly, Equation (32) shows that the mobility is:

$$u \equiv \frac{e}{kT}(D_A c_A + D_B c_B), \tag{35}$$

which, in fact, is the same as Equation (5), but now, the D_i's are mole fraction dependent. This results in a minimum conductivity value for a mole fraction close to that at which the interdiffusion coefficient reaches its maximum. In contrast, the basic model leads to monotonic u and $\bar{D}$ functions.

In order to impose the boundary conditions on ϕ, we need to relate it with the electric potential v. This can be done by the procedure followed in the above section, but the resulting expression is not so simple:

$$\phi = V + \frac{kT}{e}\int_0^{c_A} \frac{D_A c_A \frac{d\ln a_A}{dc_A} + D_B c_B \frac{d\ln a_B}{dc_A}}{D_A c_A + D_B c_B}\, dc_A. \tag{36}$$

If the MIE is caused by cation interactions ($D_i c_i = D_{ii} a_i$), this expression can be integrated for any particular dependency of the activities on the mole fraction:

$$\phi = V + \frac{kT}{e}\ln\frac{D_A c_A + D_B c_B}{D_{BB}}.$$

On the contrary, if cations behave ideally and relaxation processes are responsible for the MIE, then Equation (36) becomes:

$$\phi = V + \frac{kT}{e}\int_0^{c_A} \frac{D_A - D_B}{D_A c_A + D_B c_B}\, dc_A,$$

and the integration can only be performed once the dependencies of the diffusion coefficients on the mole fraction are known.

The starting point to impose the boundary conditions are the same as in Section 2, that is equal electrochemical potentials at both sides of the glass surface for cation species that can cross it and zero flux through it otherwise. However, the particular functions for the electrochemical potentials and flux densities must be chosen from the model of the glass behaviour.

3.4. Changes in the Solutions with Respect to the Basic Model

The main difference, with respect to the basic model, which is observed after the numerical solution of Equation (31), comes from the presence of a maximum in the interdiffusion coefficient for intermediate molar fractions (Figure 1b). Therefore, new qualitative shapes of the profiles are seen when almost all indigenous cations are replaced by the foreign ones.

In Figure 2, we show a simulation of molar fraction profiles of potassium cations in a K^+/Na^+ thermal exchange for several boundary conditions ($c_K|_S$). The profile for the lowest value of $c_K|_S$ was similar to an erfc function. For intermediate values (0.4 and 0.6), the profiles showed a bump between the glass surface and the tail, near which there was an inflection point. Both characteristics were similar to those predicted by the basic model. Nevertheless, a second inflection point (or equivalently, a high slope at the surface) appeared for the highest values of the boundary condition. Because the interdiffusion was, in the present model, lower near the surface, the slope of the profile must increase to provide cations at a sufficient rate for the intermediate regions, where they diffuse faster. An approximate analytical profile was proposed to describe all of these cases [75]. In this work, the authors checked the quality of that profile through the measurement of the effective indices of ion exchange waveguides. They obtained an average deviation of 3×10^{-4} between measured and calculated effective indices, which corresponded to a deviation of $\approx 0.3\%$ between concentration profiles. Similar results have been obtained by numerically solving the diffusion equation that governs the IE process [22]. Moreover, this author even monitored such a process as a function of the diffusion time.

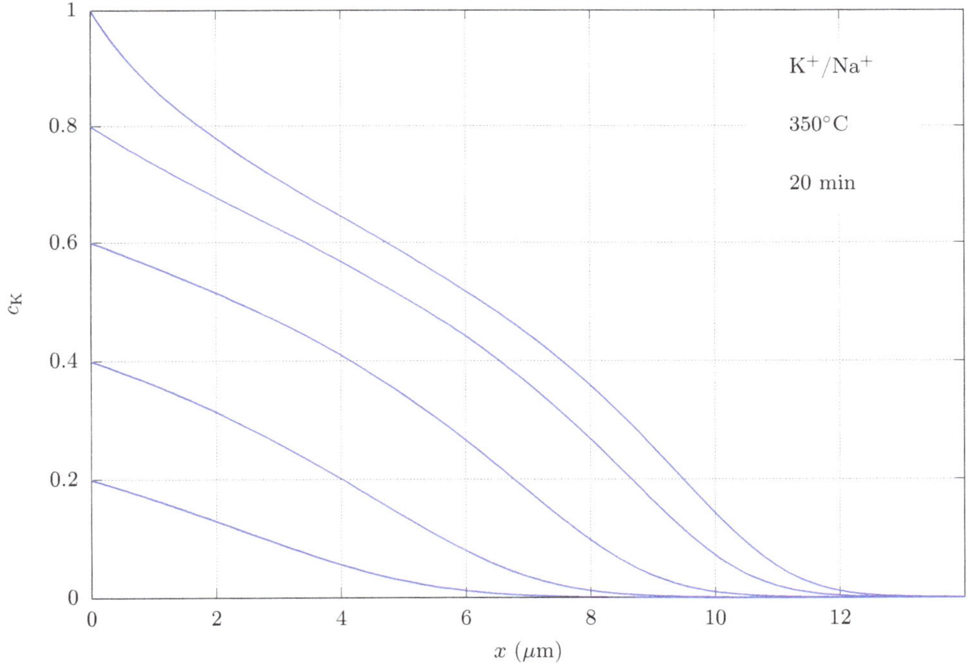

Figure 2. Simulation of molar fraction profiles as a function of depth for a thermal IE by taking into account the MIE. Equation (31) was solved for $J_0 = 0$ and the boundary conditions $c_K|_S = 0.2, 0.4, 0.6, 0.8$ and 1. The interdiffusion coefficient $\bar{D}(c_K)$ of Figure 1b was used.

A simulation for field-assisted IE can be seen in Figure 3. For the lowest boundary conditions ($c_K|_S = 0.1$–0.3), the exchange was dominated by the diffusive term, and the stable profile was not formed yet. For intermediate values of $c_K|_S$ (0.4–0.8), the stable profile was clearly formed, since the rear region of the profile was flat. Moreover, we can see that the velocity of the front and its slope increased with the boundary condition. Again, these characteristics agreed with the prediction of the basic model. Finally, for $c_K|_S = 0.9$–1, a new effect appeared. The speed and the slope of the front saturated, while its rear part was no longer flat because the potassium cations were faster here than the sodium ones. Both regimes can be observed in experimental studies [76,77].

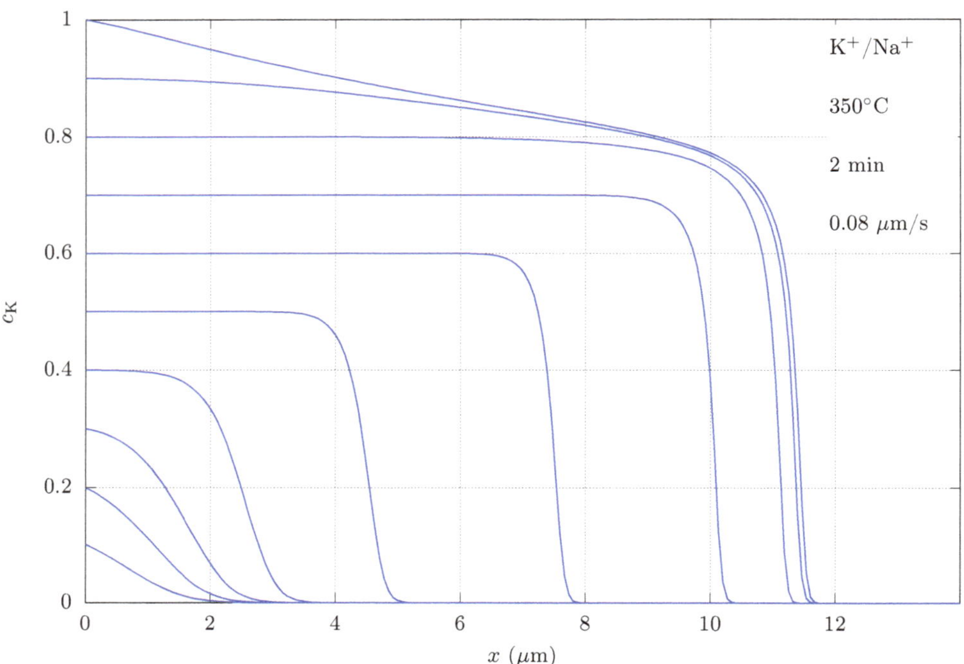

Figure 3. Simulation similar to that in Figure 2, but for a field-assisted IE with $J_0/C_0 = 80\,\mathrm{nm/s}$.

4. Future Challenges

Some of the previous assumptions in the modelling of ion exchange in glass are not always valid. For example, the interdiffusion coefficient is reduced during the ion exchange of dopants, which generates compressive stress. This is probably due to a constriction of the interstitial sites [78–80]. Another issue is the generalization of the charge neutrality approximation to processes involving the simultaneous diffusion of three species (for example sodium, silver and potassium). Some of such processes were proposed in the past to improve the shape of narrow-channel waveguides [81,82] or, more recently, to obtain both antimicrobial and strengthening properties [10]. In [82], a simplified model was presented by considering that potassium cations are immobile. Unfortunately, little progress has been made in all of these issues in the last few years. Note that modelling of these processes in the MIE framework would require considerable experimental studies to obtain the interdiffusion coefficients. Nevertheless, there are some processes that also require more complete models, but that have recently attracted a great deal of attention due to their remarkable applications. Among them, we highlight: glass poling, electro-diffusion of multivalent metals and the formation/dissolution of silver nanoparticles.

4.1. Glass Poling

Glass poling is the distribution of the electric charges of the glass. This may result in a permanent electric field inside the glass, which can give rise to relevant non-linear effects on light propagation. Applications of glass poling include grating fabrication [83], second-harmonic generation [84] or the fabrication of optical waveguides with profiled electrodes [85]. Glass poling was initially applied to silica glass [86], which contains residual amounts of cations ($C_0 \sim 10^{23}$ m^{-3}). It is realized by subjecting the sample to a strong potential difference U (typically, a few kV) at a temperature of about 250–300 °C and, usually, not allowing charged species to enter the glass ("blocking anode") [87]. Therefore, the field rearranges the cations until a new equilibrium is achieved when the original field inside the glass is cancelled by charges located near the glass surface. That is, the applied field creates a layer, under the anode, a few micrometers thick and depleted of cations. Obviously, charge neutrality is not fulfilled here. Instead, the one-dimensional form of Equations (3) and (10) of the basic model shows that the layer thickness d, in equilibrium, is given by:

$$d = \sqrt{\frac{2\epsilon U}{eC_0}} \, . \tag{37}$$

This layer is charged and has a high electrical resistivity since its anions are fixed to the glass network. Furthermore, because the field is cancelled in the glass bulk, the applied voltage drops across the depleted layer. Therefore, a very strong field arises, which is independent of the thickness of the substrate. Besides, potentials above $\sim$1 kV generate structural changes in the layer [88]. Then, if the sample is sharply cooled to room temperature, the distribution of charges becomes frozen. Electric fields inside the silica glass can reach $\sim$10^7 V/m, which induces a non-linear coefficient $\chi(2) \sim 1$ pm/V [89]. However, this process is idealized. In practice, other cations are often involved due to the high applied voltages. One of the possible cations is hydronium (H$_3$O$^+$) from air water vapour that forms a naturally hydrated layer in the glass. In this situation, called poling with a non-blocking anode, Equation (37) is no longer valid. Instead, a stable profile, such as described in Section 2.4, is formed because hydronium moves slower than sodium. In this case, a partially depleted region is still formed before the electric field increases enough to move the hydronium cations significantly [90]. The charge and the electric field are expected to be at their greatest values when the hydronium layer begins to form. Subsequently, a part of the applied voltage U drops across the hydronium layer [91]. In glasses with high alkali content, the situation is somewhat different, since air cannot provide hydronium at a sufficient rate. Moreover, as C_0 is large, d is small (see Equation (37)), and the electric field can become high enough to move other cations such as Ca^{+2} or Mg^{+2} [31,92]. Another situation of interest occurs when a BK7 glass, which contains sodium and potassium, is poled. Initially, only sodium cations are removed from the glass surface, as they move faster. Then, the field increases in the charged layer until potassium cations start to move and accumulates behind the sodium ones, filling the empty sites that the latter have just left. Simultaneously, a fully depleted surface layer is formed [92]. Poling of double-alkali glasses, with a non-blocking anode, was also simulated with similar conclusions [93]. One interesting result, which is obtained in high-alkali glasses, is the formation of a waveguide next to the depleted layer, this having non-linear properties. In all the previously cited works about the simulation of glass poling, the authors assumed diffusion coefficients constant, which is only an approximation. The diffusion coefficient in monoalkali glasses depends notably on the alkali content. Therefore, it is expected that the diffusion coefficient in the depleted layer is also different from bulk glass. Therefore, more experimental research on glass poling is still necessary to make clear, on the one hand, the conditions under which the charge neutrality approximation is valid and, on the other hand, whether the MIE or other effects are relevant enough to be considered in the modelling.

4.2. Electro-Diffusion of Multivalent Metals

Monovalent cations as Ag$^+$/Na$^+$ and K$^+$/Na$^+$ are by far the most used cations in IE processes in glass. However, some successful attempts at doping glass with multivalent cations have also been realized, mainly with transition metals, but also with rare-earths. For example, Gonella et al. obtained Co^{2+}, Au^{3+} and Cr^{3+} diffusion profiles in silicate glasses [94,95]; and Cattaruzza et al. introduced Er^{3+} cations inside soda-lime glasses [96]. These authors used field-assisted configurations with electric fields up to 400 V/mm, as thermal-assisted IE is not very effective with multivalent ion species. This allowed for penetrations of cations of up to $\sim$1 µm from deposited films. The main utility of glass doping with multivalent transition metals arises from the possibility of converting the glass into an active media. Therefore, optical amplification or waveguide lasers have been demonstrated in erbium-doped glasses [97], and Cr^{4+}-doped materials have been proposed as both laser gain materials and saturable absorbers for passive Q-switching in IR lasers [98].

The ion exchange with multivalent cations gives rise to concentration-dependent structural changes in the glass matrix. This is due to the local coordination rearrangements, which are produced at the ion sites [99]. In addition, the amount of metal that penetrates into the glass matrix, as well as the shape of the concentration profiles depend strongly on the process parameters. Therefore, the model presented in Section 3, which includes the MIE, should be used in the modelling of such processes. Likewise, when the multivalent cations are introduced into the glass, a region depleted of cations, similar to that observed in poled glasses, is formed. Therefore, the approaches used in the modelling of glass poling could be applied to the present processes. However, little work has been done in these directions, and a comprehensive model of electro-diffusion of multivalent metals is still lacking.

4.3. Formation/Dissolution of Silver Nanoparticles

Metal nanocluster formation following ion exchange can arise if both suitable dopant cations and post-exchange processing techniques are used. Among all dopants, silver stands out for its great diffusivity in glass and its high tendency to form metal clusters [33]. As for the post-exchange techniques, heat treatments (annealing) under different atmospheres [100,101] and (pulsed) laser irradiation [102] are commonly applied. The formation of metal structures can be regarded as a drawback and an advantage. For example, in light waveguiding, metal inhomogeneities must be avoided because they cause a high light absorption. On the contrary, some investigations have found these structures useful for sensing applications, through the surface enhancement of Raman scattering (SERS) [103], for fabricating photonic crystals [104] or, in general, for plasmonic optics [105,106].

Modelling of the formation and dissolution of silver nanoparticles is a very challenging task because the two complex processes compete dynamically. On the one hand, annealing or irradiation promotes silver aggregation and the formation of nanometre-sized clusters while, at the same time, the diffusion and dissolution of silver cations are produced. Therefore, at least three species must be included in the modelling: both exchanging cations and metallic silver. Moreover, this process is often done in gaseous atmospheres, which provide other cations that diffuse into the glass, increasing the number of species to be included in the model. On the other hand, this kind of process is strongly dependent on the cation concentration, so that the MIE should be considered. Finally, not enough experimental studies have been realized until now to characterize the great variety of mechanisms involved in these processes. Some theoretical descriptions of nanoparticle formation under specific post-exchange processing techniques (purely thermal annealing [107] and annealing in hydrogen atmosphere [108]) have been given. Both authors assumed the charge neutrality approximation that, under these processing techniques, is probably fulfilled. However, they did not assess the introduction of the MIE in their models and if that would lead to a better description of the cluster formation. Considerable work, both theoretical and experimental, must be done before having a complete theoretical model of this promising technique.

5. Conclusions

The improvements made in the last few years in the theoretical modelling of ion exchange in glass have contributed to a better understanding of the physical and chemical mechanisms involved in this process. The basic model that has been used for decades only allows for the accurate description of ion exchange problems where the dopant concentration is low. If higher concentrations are present, a model based on the mixed alkali effect must be considered. However, some aspects of this model are still open, because the physics and chemistry involved are not fully understood. On the other hand, a significant progress has been made in the modelling of some IE processes that have attracted much attention in recent years. These processes include glass poling, electro-diffusion of multivalent metals, and the formation/dissolution of silver nanoparticles. All of them have remarkable applications, both scientific, as well as technological.

Funding: This research was funded by Xunta de Galicia, Consellería de Educación, Universidades e FP, Grant GRC Number ED431C2018/11, and Ministerio de Economía, Industria y Competitividad, Gobierno de España, Grant Number AYA2016-78773-C2-2-P, European Regional Development Fund.

Institutional Review Board Statement: Not applicable.

Informed Consent Statement: Not applicable.

Data Availability Statement: Data used in this article data were extracted from Fleming, J.W., Jr.; Day, D.E. Relation of Alkali Mobility and Mechanical Relaxation in Mixed-Alkali Silicate Glasses. *J. Am. Ceram. Soc.* **1972**, *55*, 186–192.

Conflicts of Interest: The authors declare no conflict of interest.

Abbreviations

The following abbreviations are used in this manuscript:

DC	Direct current
IE	Ion exchange
MIE	Mixed ion effect

References

1. Mazzoldi, P.; Sada, C. A trip in the history and evolution of ion exchange process. *Mater. Sci. Eng. B* **2008**, *149*, 112–117. [CrossRef]
2. Kistler, S.S. Stresses in Glass Produced by Nonuniform Exchange of Monovalent Ions. *J. Am. Ceram. Soc.* **1962**, *45*, 59–68. [CrossRef]
3. Nordberg, M.E.; Mochel, E.L.; Garfinkel, H.M.; Olcott, J.S. Strengthening by Ion Exchange. *J. Am. Ceram. Soc.* **1964**, *47*, 215–219. [CrossRef]
4. Miller, S.E. Integrated Optics: An Introduction. *Bell Syst. Tech. J.* **1969**, *48*, 2059–2069. [CrossRef]
5. Izawa, T.; Nakagone, H. Optical Waveguides Formed by Electrically Induced Migration of Ions in Glass Plates. *Appl. Phys. Lett.* **1972**, *21*, 584–586. [CrossRef]
6. Almeida, R.M.; Marques, A.C. The potential of ion exchange in sol-gel derived photonic materials and structures. *Mater. Sci. Eng. B* **2008**, *149*, 118–122. [CrossRef]
7. Gy, R. Ion exchange for glass strengthening. *Mater. Sci. Eng. B* **2008**, *149*, 159–165. [CrossRef]
8. Karlsson, S.; Jonson, B.; Stålhandske, C. The technology of chemical glass strengthening—A review. *Glass Technol.* **2010**, *51*, 41–54.
9. Varshneya, A.K. Chemical Strengthening of Glass: Lessons Learned and Yet To Be Learned. *Int. J. Appl. Glass Sci.* **2010**, *1*, 131–142. [CrossRef]
10. Guldiren, D.; Erdem, I.; Aydin, S. Influence of silver and potassium ion exchange on physical and mechanical properties of soda lime glass. *J. Non-Cryst. Solids* **2016**, *441*, 1–9. [CrossRef]
11. Macrelli, G.; Varshneya, A.K.; Mauro, J.C. Ion Exchange in Silicate Glasses: Physics of Ion Concentration, Residual Stress, and Refractive Index Profiles. *arXiv* **2020**, arXiv:2002.08016.
12. Najafi, S.I. (Ed.) *Introduction to Glass Integrated Optics*; Artech House: Boston, MA, USA; London, UK, 1992.
13. Opilski, A.; Rogoziński, R.; Gut, K.; Błahut, M.; Opilski, Z. Present state and perspectives involving application of ion exchange in glass. *Opto-Electron. Rev.* **2000**, *8*, 117–127.
14. Honkanen, S.; West, B.R.; Yliniemi, S.; Madasamy, P.; Morrell, M.; Auxier, J.; Schülzgen, A.; Peyghambarian, N.; Carriere, J.; Frantz, J.; et al. Recent advances in ion exchanged glass waveguides and devices. *Phys. Chem. Glas. Eur. J. Glass Sci. Andtechnol. Part B* **2006**, *47*, 110–120.
15. Quaranta, A.; Cattaruzza, E.; Gonella, F. Modelling the ion exchange process in glass: Phenomenological approaches and perspectives. *Mater. Sci. Eng. B* **2008**, *149*, 133–139. [CrossRef]
16. Tervonen, A.; West, B.R.; Honkanen, S. Ion exchanged glass waveguide technology: A review. *Opt. Eng.* **2011**, *50*. [CrossRef]

17. Brusberg, L.; Schröder, H.; Herbst, C.; Frey, C.; Fiebig, C.; Zakharian, A.; Kuchinsky, S.; Liu, X.; Fortusini, D.; Evans, A. High performance ion exchanged integrated waveguides in thin glass for board-level multimode optical interconnects. In Proceedings of the 2015 European Conference on Optical Communication (ECOC), Valencia, Spain, 27 September–1 October 2015; pp. 1–3. [CrossRef]

18. Wang, F.; Chen, B.; Pun, E.Y.B.; Lin, H. Alkaline aluminum phosphate glasses for thermal ion exchanged optical waveguide. *Opt. Mater.* **2015**, *42*, 484–490. [CrossRef]

19. Salmio, R.P.; Saarinen, J. Graded-Index Diffractive Elements by Thermal Ion Exchange in Glass. *Appl. Phys. Lett.* **1995**, *66*, 917–9. [CrossRef]

20. Singer, W.; Dobler, B.; Schreiber, H.; Brenner, K.H.; Messerschmidt, B. Refractive-index measurement of gradient-index microlenses by diffraction tomography. *Appl. Opt.* **1996**, *35*, 2167–2171. [CrossRef] [PubMed]

21. Montero-Orille, C.; Moreno, V.; Prieto-Blanco, X.; Mateo, E.F.; Ip, E.; Crespo, J.; Liñares, J. Ion exchanged glass binary phase plates for mode-division multiplexing. *Appl. Opt.* **2013**, *52*, 2332–2339. [CrossRef] [PubMed]

22. Rogoziński, R. Ion Exchange in Glass—The Changes of Glass Refraction. *Ion Exch. Technol.* **2012**, 155–190. [CrossRef]

23. Helfferich, F.; Plesset, M.S. Ion Exchange Kinetics. A Non Linear Diffusion Problem. *J. Chem. Phys.* **1958**, *28*, 418–424. [CrossRef]

24. Abou-el-Leil, M.; Cooper, A.R. Analysis of Field-Assisted Binary Ion Exchange. *J. Am. Ceram. Soc.* **1979**, *62*, 390–395. [CrossRef]

25. Tervonen, A. A General Model for Fabrication Processes of Channel Waveguides by Ion Exchange. *J. Appl. Phys.* **1990**, *67*, 2746–2752. [CrossRef]

26. Cheng, D.; Saarinen, J.; Saarikoski, H.; Tervonen, A. Simulation of Field-assisted Ion exchange for Glass Channel Waveguide Fabrication: Effect of Nonhomogeneous Time-dependent Electric Conductivity. *Opt. Commun.* **1997**, *134*, 233–238. [CrossRef]

27. Fleming, J.W., Jr.; Day, D.E. Relation of Alkali Mobility and Mechanical Relaxation in Mixed-Alkali Silicate Glasses. *J. Am. Ceram. Soc.* **1972**, *55*, 186–192. [CrossRef]

28. Inman, J.M.; Houde-Walter, S.; McIntyre, B.L.; Liao, Z.M.; Parker, R.S.; Simmons, V. Chemical Structure and the Mixed Mobile Ion Effect in Silver-for-Sodium Ion Exchange in Silicate Glass. *J. Non-Cryst. Solids* **1996**, *194*, 85–92. [CrossRef]

29. Lupascu, A.; Kevorkian, A.; Bondet, T.; Saint-André, F.; Persegol, D.; Levy, M. Modeling Ion Exchange in Glass with Concentration Dependent Diffusion Coefficients and Mobilities. *Opt. Eng.* **1996**, *35*, 1603–1610. [CrossRef]

30. Prieto-Blanco, X. Electro-diffusion equations of monovalent cations in glass under charge neutrality approximation for optical waveguide fabrication. *Opt. Mater.* **2008**, *31*, 418–428. [CrossRef]

31. Petrov, M.I.; Lepen'kin, Y.A.; Lipovskii, A.A. Polarization of glass containing fast and slow ions. *J. Appl. Phys.* **2012**, *112*, 043101. [CrossRef]

32. Okorn, B.; Sancho-Parramon, J.; Oljaca, M.; Janicki, V. Metal doping of dielectric thin layers by electric field assisted film dissolution. *J. Non-Cryst. Solids* **2021**, *554*, 120584. [CrossRef]

33. Gonella, F. Silver doping of glasses. *Ceram. Int.* **2015**, *41*, 6693–6701. [CrossRef]

34. Doremus, R.H. Exchange and Diffusion of Ions in Glass. *J. Phys. Chem.* **1964**, *68*, 2212–2218. [CrossRef]

35. Mrozek, P. Numerical modeling of field-assisted Ag+–Na+ ion exchanged channel waveguides using varied explicit space charge density approach. *Opt. Appl.* **2019**, *49*. [CrossRef]

36. Prieto, X.; Srivastava, R.; Liñares, J.; Montero, C. Prediction of Space-Charge Density and Space-Charge Field in Thermally Ion Exchanged Planar Surface Waveguides. *Opt. Mater.* **1996**, *5*, 145–151. [CrossRef]

37. Day, D.E. Mixed Alkali Glasses—Their Properties and Uses. *J. Non-Cryst. Solids* **1976**, *21*, 343–372. [CrossRef]

38. Batchelor, S.; Oven, R.; Ashworth, D.G. Characterization of Electric Field Assisted Diffused Potassium Ion Planar Optical Waveguides. *Electron. Lett.* **1996**, *32*, 2082–2083. [CrossRef]

39. Zheng, W.; Yang, B.; Hao, Y.; Wang, M.; Jiang, X.; Yang, J. Charge-density flux model for electric-field-assisted ion exchange in glass. *Opt. Eng.* **2011**, *50*, 1–9. [CrossRef]

40. Rothmund, V.; Kornfeld, G. Der Basenanstausch im Permutit, I. *Z. Anorg. Allg. Chem.* **1918**, *103*, 129. [CrossRef]

41. Eisenman, G. Cation selective glass electrodes and their mode of operation. *Biophys. J.* **1962**, *2*, 259–323. [CrossRef]

42. Garfinkel, H.M. Ion Exchange Equilibria between Glass and Molten Salts. *J. Phys. Chem.* **1968**, *72*, 4175–4181. [CrossRef]

43. Chludzinski, P.; Ramaswamy, R.V.; Anderson, T.J. Ion Exchange Between Soda-Lime-Silica Glass and Sodium Nitrate-Silver Nitrate Moten Salts. *Phys. Chem. Glas.* **1987**, *28*, 169–173.

44. Crank, J. *The Mathematics of Diffusion*, 2nd ed.; Oxford Clarenton Press: Oxford, UK, 1979.

45. Stewart, G.; Laybourn, P.J. Fabrication of Ion Exchanged Optical Waveguides from Dilute Silver Nitrate Melts. *IEEE J. Quantum Electron.* **1978**, *QE-14*, 930–934. [CrossRef]

46. Liñares, J.; Prieto, X.; Montero, C. A Novel Refractive Index Profile for Optical Characterization of Nonlinear Diffusion Processes and Planar Waveguides in Glass. *Opt. Mater.* **1994**, *3*, 229–236. [CrossRef]

47. Prieto, X.; Linares, J.; Montero, C. Perturbative method to modelize ion exchange processes: Application to surface waveguides. In *Fiber and Integrated Optics*; International Society for Optics and Photonics: Bellingham, WA, USA, 1996; Volume 2954, pp. 62–66.

48. Prieto, X.; Liñares, J. Increasing Resistivity Effects in Field-Assisted Ion Exchange for Planar Optical Waveguide Fabrication. *Opt. Lett.* **1996**, *21*, 1–3. [CrossRef]

49. Anderson, O.L.; Stuart, D.A. Calculation of Activation Energy of Ionic Conductivity in Silica Glasses by Classical Methods. *J. Am. Ceram. Soc.* **1954**, *37*, 573–580. [CrossRef]

50. McVay, G.L.; Day, D.E. Diffusion and Internal Friction in Na-Rb Silicate Glasses. *J. Am. Ceram. Soc.* **1970**, *53*, 508–513. [CrossRef]
51. Hayami, R.; Terai, R. Diffusion of Alkali Ions in Na_2O-Cs_2O-SiO_2 Glasses. *Phys. Chem. Glas.* **1972**, *13*, 102–106.
52. Isard, J. The mixed alkali effect in glass. *J. Non-Cryst. Solids* **1969**, *1*, 235–261. [CrossRef]
53. Terai, R.; Hayami, R. Ionic Diffusion in Glasses. *J. Non-Cryst. Solids* **1975**, *18*, 217–264. [CrossRef]
54. Ingram, M.D. Ionic conductivity in glass. *Phys. Chem. Glas.* **1987**, *28*, 215–234.
55. Araujo, R.J. Interdiffusion in a One-Dimensional Interacting System. *J. Non-Cryst. Solids* **1993**, *152*, 70–74. [CrossRef]
56. Inman, J.M.; Bentley, J.L.; Houde-Walter, S. Modeling Ion Exchanged Glass Photonics: The Modified Quasi-Chemical Diffusion Coefficient. *J. Non-Cryst. Solids* **1995**, *191*, 209–215. [CrossRef]
57. Tomozawa, M. Structure of mixed alkali glasses. *J. Non-Cryst. Solids* **1996**, *196*, 280–284. [CrossRef]
58. Kahnt, H. Ionic transport in glasses. *J. Non-Cryst. Solids* **1996**, *203*, 225–231. [CrossRef]
59. Ngai, K. The dynamics of ions in glasses: Importance of ion–ion interactions. *J. Non-Cryst. Solids* **2003**, *323*, 120–126. [CrossRef]
60. Terai, R. The Mixed Alkali Effect in the Na_2O-Cs_2O-SiO_2 Glasses. *J. Non-Cryst. Solids* **1971**, *6*, 121–135. [CrossRef]
61. Terai, R.; Wakabayashi, H.; Yamanaka, H. Haven ratio in mixed alkali glass. *J. Non-Cryst. Solids* **1988**, *103*, 137–142. [CrossRef]
62. Messerschmidt, B.; Hsieh, C.H.; McIntyre, B.L.; Houde-Walker, S. Ionic Mobility in an Ion Exchanged Silver-Sodium Boroaluminosilicate Glass for Micro-Optics Applications. *J. Non-Cryst. Solids* **1997**, *217*, 264–271. [CrossRef]
63. Lezzi, P.; Tomozawa, M. Enthalpy of mixing of mixed alkali glasses. *J. Non-Cryst. Solids* **2010**, *356*, 1439–1446. [CrossRef]
64. Lezzi, P.; Tomozawa, M. Effect of alumina on enthalpy of mixing of mixed alkali silicate glasses. *J. Non-Cryst. Solids* **2011**, *357*, 2086–2092. [CrossRef]
65. Kouyate, A.; Ahoussou, A.; Yapi, A.; Diabate, D.; Rogez, J.; Trokourey, A. Influence of alkali mixed effect on the mixing enthalpy in $0.75\,B_2O_3 - 0.25\,[xNa_2O$–$(1-x)K_2O]$ glass system. *Chin. Chem. Lett.* **2008**, *19*, 1252–1255. [CrossRef]
66. Florian, P.; Vermillion, K.; Grandinetti, P.; Farnan, I.; Stebbins, J. Cation distribution in mixed alkali disilicate glasses. *J. Am. Chem. Soc.* **1996**, *118*, 3493–2497. [CrossRef]
67. Gee, B.; Eckert, H. Cation Distribution in Mixed-Alkali Silicate Glasses. NMR Studies by ^{23}Na-{^{7}Li} and ^{23}Na-{^{6}Li} Spin Echo Double Resonance. *J. Phys. Chem.* **1996**, *100*, 3705–3712. [CrossRef]
68. Ratai, E.; Chan, J.C.; Eckert, H. Local coordination and spatial distribution of cations in mixed-alkali borate glasses. *Phys. Chem. Chem. Phys.* **2002**, *4*, 3198–3208. [CrossRef]
69. Swenson, J.; Matic, A.; Karlsson, C.; Börjesson, L.; Meneghini, C.; Howells, W. Random ion distribution model: A structural approach to the mixed-alkali effect in glasses. *Phys. Rev. B* **2001**, *63*, 132202. [CrossRef]
70. Maass, P.; Bunde, A.; Ingram, M.D. Ion transport anomalies in glasses. *Phys. Rev. Lett.* **1992**, *68*, 3064–3067. [CrossRef]
71. Bunde, A.; Ingram, M.D.; Maass, P. The Dynamic Structure Model for Ion Transport in Glasses. *J. Non-Cryst. Solids* **1994**, *172–174*, 1222–1236. [CrossRef]
72. Bunde, A.; Funke, K.; Ingram, M.D. A unified site relaxation model for ion mobility in glassy materials. *Solid State Ionics* **1996**, *86*, 1311–1317. [CrossRef]
73. Davidson, J.E.; Ingram, M.D.; Bunde, A.; Funke, K. Ion hopping processes and structural relaxation in glassy materials. *J. Non-Cryst. Solids* **1996**, *203*, 246–251. [CrossRef]
74. Murch, G.E. The Nernst-Einstein equation in high-defect-content solids. *Philos. Mag. A* **1982**, *45*, 685–692. [CrossRef]
75. Prieto-Blanco, X.; Montero, C.; Crespo, J.; Barral, D.; Mouriz, D.; Nistal, M.C.; Mateo, E.F.; Moreno, V.; Liñares, J. Hyperbolic interdiffusion for the double alkali effect on index profiles of ion exchanged glass slab optical elements. In Proceedings of the 3rd Congress of the International Commission for Optics, Santiago de Compostela, Spain, 26–29 August 2014.
76. Prieto Blanco, X. Interferometric characterization and analysis of silver-exchanged glass waveguides buried by electromigration: slab, channel and slab-sided channel configurations. *J. Opt. A Pure Appl. Opt.* **2006**, *8*, 123. [CrossRef]
77. Sviridov, S.I. Influence of the composition of a molten salt on the field-assisted diffusion of potassium ions in the $20Na_2O\cdot80SiO_2$ glass. *Glass Phys. Chem.* **2007**, *33*, 550–555. [CrossRef]
78. Nogami, M.; Tomozawa, M. Effect of Stress on Water Diffusion in Silica Glass. *J. Am. Ceram. Soc.* **1984**, *67*, 151–154. [CrossRef]
79. Varshneya, A.K.; Dumais, G.A. Influence of Externally Applied Stresses on Kinetics of Ion Exchange in Glass. *J. Am. Ceram. Soc.* **1985**, *68*, C-165–C-166. [CrossRef]
80. Varshneya, A.K. The physics of chemical strengthening of glass: Room for a new view. *J. Non-Cryst. Solids* **2010**, *356*, 2289–2294. [CrossRef]
81. Cooper, A.R. Method for Fabricating Buried Waveguides. EP 0 380 468 A2. Available online: https://data.epo.org/publication-server/document?iDocId=660504&iFormat=0 (accessed on 24 April 2008).
82. Tervonen, A.; Honkanen, S. Model for Waveguide Fabrication in Glass by Two-Step Ion Exchange with Ionic Masking. *Opt. Lett.* **1988**, *13*, 71–73. [CrossRef] [PubMed]
83. Lipovskii, A.A.; Rusan, V.V.; Tagantsev, D.K. Imprinting phase/amplitude patterns in glasses with thermal poling. *Solid State Ionics* **2010**, *181*, 849–855. [CrossRef]
84. Garcia, F.C.; Carvalho, I.C.S.; Hering, E.; Margulis, W.; Lesche, B. Inducing a large second-order optical nonlinearity in soft glasses by poling. *Appl. Phys. Lett.* **1998**, *72*, 3252–3254. [CrossRef]
85. Zhurikhina, V.; Sadrieva, Z.; Lipovskii, A. Single-mode channel optical waveguides formed by the glass poling. *Optik* **2017**, *137*, 203–208. [CrossRef]

86. Myers, R.A.; Mukherjee, N.; Brueck, S.R.J. Large second-order nonlinearity in poled fused silica. *Opt. Lett.* **1991**, *16*, 1732–1734. [CrossRef]

87. Fleming, S.C.; An, H. Poled glasses and poled fibre devices. *J. Ceram. Soc. Jpn.* **2008**, *116*, 1007–1023. [CrossRef]

88. An, H.; Fleming, S. Near-anode phase separation in thermally poled soda lime glass. *Appl. Phys. Lett.* **2006**, *88*, 181106. [CrossRef]

89. Kudlinski, A.; Quiquempois, Y.; Martinelli, G. Modeling of the $\chi(2)$ susceptibility time-evolution in thermally poled fused silica. *Opt. Express* **2005**, *13*, 8015–8024. [CrossRef]

90. Petrov, M.I.; Omelchenko, A.V.; Lipovskii, A.A. Electric field and spatial charge formation in glasses and glassy nanocomposites. *J. Appl. Phys.* **2011**, *109*, 094108. [CrossRef]

91. Lipovskii, A.; Omelchenko, A.; Petrov, M. Modeling charge transfer dynamics and electric field distribution in glasses during poling and electrostimulated diffusion. *Tech. Phys. Lett.* **2010**, *36*, 1028–1031. [CrossRef]

92. Brennand, A.L.R.; Wilkinson, J.S. Planar waveguides in multicomponent glasses fabricated by field-driven differential drift of cations. *Opt. Lett.* **2002**, *27*, 906–908. [CrossRef] [PubMed]

93. Oven, R. Analytical model of electric field assisted ion diffusion into glass containing two indigenous mobile species, with application to poling. *J. Non-Cryst. Solids* **2021**, *553*, 120476. [CrossRef]

94. Gonella, F.; Cattaruzza, E.; Quaranta, A.; Ali, S.; Argiolas, N.; Sada, C. Diffusion behavior of transition metals in field-assisted ion exchanged glasses. *Solid State Ionics* **2006**, *177*, 3151–3155. [CrossRef]

95. Cattaruzza, E.; Battaglin, G.; Gonella, F.; Quaranta, A.; Mariotto, G.; Sada, C.; Ali, S. Chromium doping of silicate glasses by field-assisted solid-state ion exchange. *J. Non-Cryst. Solids* **2011**, *357*, 1846–1850. [CrossRef]

96. Cattaruzza, E.; Gonella, F.; Peruzzo, G.; Quaranta, A.; Sada, C.; Trave, E. Field-assisted ion diffusion in dielectric matrices: Er^{3+} in silicate glass. *Mater. Sci. Eng. B* **2008**, *146*, 163–166. [CrossRef]

97. Kenyon, A. Recent developments in rare-earth doped materials for optoelectronics. *Prog. Quantum Electron.* **2002**, *26*, 225–284. [CrossRef]

98. Kalisky, Y. Cr4+-doped crystals: their use as lasers and passive Q-switches. *Prog. Quantum Electron.* **2004**, *28*, 249–303. [CrossRef]

99. Thévenin-Annequin, C.; Levy, M.; Pagnier, T. Electrochemical study of the silver-sodium substitution in a borosilicate glass. *Solid State Ionics* **1995**, *80*, 175–179. [CrossRef]

100. Marchi, G.D.; Caccavale, F.; Gonella, F.; Mattei, G.; Mazzoldi, P.; Battaglin, G.; Quaranta, A. Silver nanoclusters formation in ion exchanged waveguides by annealing in hydrogen atmosphere. *Appl. Phys. A* **1996**, *63*, 403–407. [CrossRef]

101. Mohr, C.; Dubiel, M.; Hofmeister, H. Formation of silver particles and periodic precipitate layers in silicate glass induced by thermally assisted hydrogen permeation. *J. Phys. Condens. Matter* **2000**, *13*, 525–536. [CrossRef]

102. Cattaruzza, E.; Mardegan, M.; Trave, E.; Battaglin, G.; Calvelli, P.; Enrichi, F.; Gonella, F. Modifications in silver-doped silicate glasses induced by ns laser beams. *Appl. Surf. Sci.* **2011**, *257*, 5434–5438. [CrossRef]

103. Chen, Y.; Karvonen, L.; Säynätjoki, A.; Ye, C.; Tervonen, A.; Honkanen, S. Ag nanoparticles embedded in glass by two-step ion exchange and their SERS application. *Opt. Mater. Express* **2011**, *1*, 164–172. [CrossRef]

104. Kaganovsky, Y.S.; Antonov, I.; Rosenbluh, M.; Ihlemann, J.; Lipovskii, A. Two- and Three-Dimensional Photonic Crystals Produced by Pulsed Laser Irradiation in Silver-Doped Glass. Functional Nanomaterials for Optoelectronics and other Applications. In *Solid State Phenomena*; Trans Tech Publications Ltd.: Stafa-Zurich, Switzerland, 2004; Volume 99, pp. 65–72. [CrossRef]

105. Kumar, P.; Mathpal, M.C.; Tripathi, A.K.; Prakash, J.; Agarwal, A.; Ahmad, M.M.; Swart, H.C. Plasmonic resonance of Ag nanoclusters diffused in soda-lime glasses. *Phys. Chem. Chem. Phys.* **2015**, *17*, 8596–8603. [CrossRef]

106. Warren, S.; Thimsen, E. Plasmonic solar water splitting. *Energy Environ. Sci.* **2012**, *5*, 5133–5146. [CrossRef]

107. Berger, A. Concentration and size depth profile of colloidal silver particles in glass surfaces produced by sodium-silver ion exchange. *J. Non-Cryst. Solids* **1992**, *151*, 88–94. [CrossRef]

108. Redkov, A.; Zhurikhina, V.; Lipovskii, A. Formation and self-arrangement of silver nanoparticles in glass via annealing in hydrogen: The model. *J. Non-Cryst. Solids* **2013**, *376*, 152–157. [CrossRef]

Review

Integrated Photonics on Glass: A Review of the Ion-Exchange Technology Achievements

Jean-Emmanuel Broquin [1,*] and Seppo Honkanen [2]

[1] IMEP-LAHC, University Savoie Mont Blanc, University Grenoble Alpes, CNRS, Grenoble INP, 38000 Grenoble, France

[2] Institute of Photonics, University of Eastern Finland, 80100 Joensuu, Finland; seppo.honkanen@uef.fi

[*] Correspondence: jean-emmanuel.broquin@grenoble-inp.fr

Featured Application: ion-exchange on glass has been extensively studied for the realization of Planar Lightwave Circuits. Monolithically integrated on a single glass wafer, these devices have been successfully employed in optical communication systems as well as in sensing.

Abstract: Ion-exchange on glass is one of the major technological platforms that are available to manufacture low-cost, high performance Planar Lightwave Circuits (PLC). In this paper, the principle of ion-exchanged waveguide realization is presented. Then a review of the main achievements observed over the last 30 years will be given. The focus is first made on devices for telecommunications (passive and active ones) before the application of ion-exchanged waveguides to sensors is addressed.

Keywords: integrated photonics; glass photonics; optical sensors; waveguides; lasers

Citation: Broquin, J.-E.; Honkanen, S. Integrated Photonics on Glass: A Review of the Ion-Exchange Technology Achievements. *Appl. Sci.* **2021**, *11*, 4472. https://doi.org/10.3390/app11104472

Academic Editor: Alessandro Belardini

Received: 25 April 2021
Accepted: 11 May 2021
Published: 14 May 2021

Publisher's Note: MDPI stays neutral with regard to jurisdictional claims in published maps and institutional affiliations.

1. Introduction

Unlike microelectronics where the CMOS technology emerged as the dominant platform, integrated optics or, as it is called nowadays, integrated photonics, does not rely on one single technological platform. Indeed, silicon photonics, III-V photonics, polymer photonics, LiNbO$_3$ photonics, and, last but not least, glass photonics co-exist in parallel, each of them presenting their own drawbacks and advantages.

As for ion-exchange on glass, also called glass integrated optics, it is based on a material that has been known and used for centuries. Glass is easily available and can be easily recycled. The ion-exchange technique, although it is based on using microfabrication tools, can be considered as a relatively low-cost approach, which allows realizing waveguides with low propagation losses and a high compatibility with optical fibers. Glass photonics is not a platform that has been developed for a specific application. Therefore, Planar Lightwave Circuits (PLCs) realized by ion-exchange on glass are found in many fields with a wide range of applications.

From its very beginning in 1972 [1], to products currently on the markets, thousands of papers have been published on this vivid topic. For this reason, making an extensive review of this technology is a cumbersome task. However, since excellent reviews have already been published in the past years [2–9], we can skip the pioneering years when the basis of the technology was set by testing several glasses and ions and making multimode waveguides. In this paper, we will hence focus on devices made by ion-exchange on glass, their performances, and their applications.

After a presentation of ion-exchanged waveguides, their realization process, their modelling, and their main characteristics, we will review devices made for telecommunication purpose. Then, we will review the use of ion-exchanged waveguides for the fabrication of optical sensors since these types of applications are taking a growing place in integrated photonics.

2. Ion-Exchanged Waveguides

2.1. Principle and Technology

Typically, an optical glass is an amorphous material composed by several types of oxides mixed together. According to Zachariasen [10], theses oxides can be sorted in three main categories: network formers like SiO_2, GeO_2, or P_2O_5 that can create a glass on their own; intermediate network formers (Al_2O_3, TiO_2, ...) that can hardly create a glass alone but can be combined with network formers; finally network modifier oxides like Na_2O, K_2O, CaO, or BaO that can be inserted in a matrix made by glass formers but are weakly linked to it because of a mismatch between their respective molecular binding structures.

The refractive index of a glass depends on its composition through an empirical relation [11]:

$$n = 1 + \sum_m \frac{a_m N_m}{V_0} = 1 + \frac{R_0}{V_0},$$ (1)

where a_m is the "refractivity constant" of the chemical element "m", N_m the number of chemical element "m" by atom of oxygen, V_0 and R_0 are the glass volume and refractivity by atom of oxygen, respectively.

A replacement of a portion of one of the glass components by another one with the same coordination can therefore entail a change of refractive index. Providing that this exchange does not create strong mechanical stresses and does not strongly change the nature of the glass, (1) can be used to link the induced variation of the refractive index to the fraction c of substituting ions as follows:

$$\Delta n = \frac{c}{V_0} \left(\Delta R - \frac{\Delta V\, R_0}{V_0} \right),$$ (2)

ΔR and ΔV are the variation of R_0 and V_0, respectively, caused by the substitution. From (2), it can easily be deduced that a local change of the glass composition is creating a localized change of refractive index, which can be used to create a waveguide.

Since alkali ions are weakly linked to the glass matrix, they are natural candidates for such a process. Indeed, when alkali ions react with silica to form a multicomponent glass, the silica network is maintained because each silicon-oxygen tetrahedron remains linked to at least three other tetrahedra [12]. Therefore, one can exchange one alkali ion to another one without damaging the original glass. Throughout the years, several ion-exchanges have been demonstrated [13,14] but the topic of this article being integrated glass photonics, we will restrain ourselves on the few ones that have enabled realizing efficient devices. In this case, the ion that is present in the glass is usually Na^+ (sometimes K^+). It is nowadays mostly exchanged with silver (Ag^+), more rarely with potassium (K^+) or thallium (Tl^+).

The ion source that allows creating the higher refractive index waveguide's core can be either liquid or solid. The simplest way of performing an ion-exchange is described on Figure 1a. It consists in dipping the glass wafer in a molten salt containing a mixture of both the doping ions B^+ and the glass ones A^+. The salt is usually a nitrate, but sulfates are sometimes used when a temperature higher than 450 °C is required for the exchange. Although the principle of the process is very simple, it must be kept in mind that ionic diffusion is a process that strongly depends on the temperature; this parameter should hence be homogeneous all other the wafer and consequently in the molten salt. In order to define the parts of the wafer that will be ion-exchanged, a thin-film has previously been deposited and patterned in a clean room environment to define the diffusion apertures. Once the ion-exchange is completed, the masking layer is removed and diffused surface waveguides are obtained. If a more step-like refractive index profile is required, an electric field can be applied to push the doping ions inside the glass, as described in Figure 1b [1]. Nonetheless, this complicates the set-up and might also induce the reduction of the doping ions into metallic clusters that dramatically increase the propagation losses (specifically when silver is involved). The use of a silver thin film has also been employed successfully for the creation of the waveguide's core [15]. The thin film can be either deposited on an existing mask, as depicted on Figure 1c, or patterned directly on the glass substrate [16]. An applied electric field ensures an efficient electrolysis of Ag^+ ions into the glass by the consumption of the silver film anode. These three different processes allow realizing waveguides whose core is placed at the surface of the glass wafer and whose shape is, depending on the process parameters, semi-elliptical with a step refractive index change at their surface and diffused interfaces inside the glass. Intrinsically, such waveguides are supporting modes that are prone to interact with the elements present on the wafer surface. Interesting and even maximized for the realization of sensors, this interaction is often a drawback when dealing with telecom devices where the preservation of the quality of the optical signal is a key factor. For this reason, ion-exchanged waveguide cores are usually buried inside the glass.

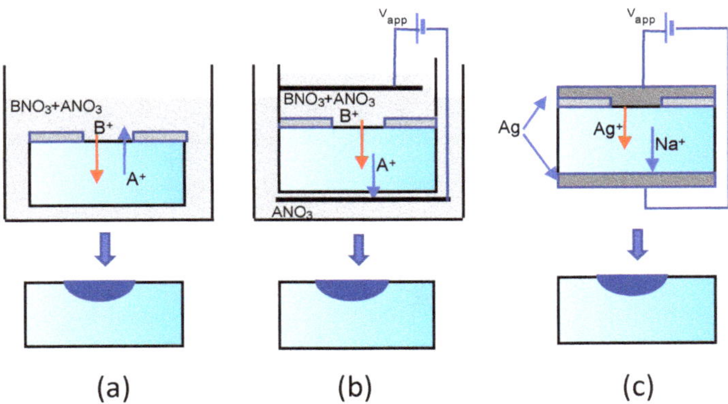

Figure 1. Three main processes used to realize surface waveguides by an ion-exchange on glass. A^+ and B^+ represent the ions contained in the glass and the ones replacing them, respectively. (**a**) the glass wafer is dipped into a molten salt containing B^+ ions entailing a thermal diffusion on the exchange ions through a diffusion aperture; (**b**) the diffusion process is assisted by an electric field; (**c**) an electrolysis of a silver thin film is used to generate Ag^+ ions that are migrating by diffusion and conduction inside the glass.

Figure 2 depicts the two main processes that can be used: the first one consists of plunging the wafer containing surface cores in a molten salt containing only the ions that were originally present in the glass. A reverse ion-exchange is then occurring, removing doping ions from the surface of the glass [17]. This process entails a quite important decrease of the refractive index change and an increase of the waveguide's dimension because of thermal diffusion, which practically limits the depth of the burying to one to two micrometers. In order to reach a deeper depth and ensure a good optical insulation of the guided mode, the reverse ion-exchange is quite often assisted by an electric field that forces the migration of the core inside the glass preventing hence a loss of refractive index variation. Moreover, by a proper tuning of the process parameters, circular waveguide cores can be obtained in order to maximize the coupling efficiency with optical fibers. Nonetheless, it must be noticed that the applied voltage can be close to 1 kV, which requires on one hand, a proper and well secured dedicated set-up, and on the other hand, an excellent quality of the glass wafer in order to prevent percolation path formation and short circuits. Figure 3 depicts an optical image of a buried optical waveguide realized on a Teem Photonics GO14 glass by a silver-sodium ion-exchange. Burying depth as high as 47 µm have been realized, as shown in Figure 4, but such extreme values are rarely required in practical devices where the burying depth is of the order of 10 µm.

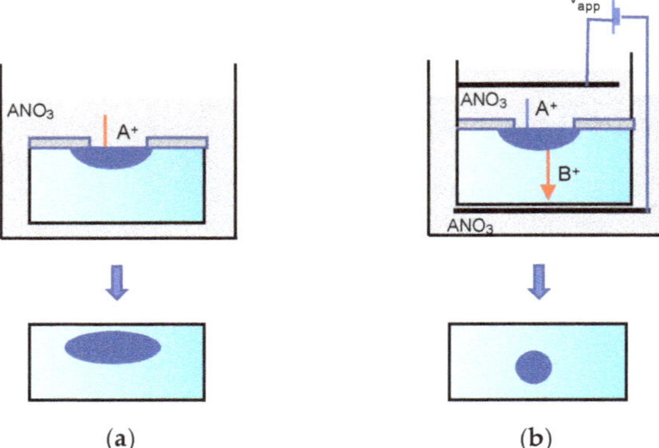

Figure 2. (**a**) Thermal burying of a waveguide's core; (**b**) electrically assisted burying of the waveguide's core. The competition between ionic diffusion and transport allows obtaining quasi circular profiles.

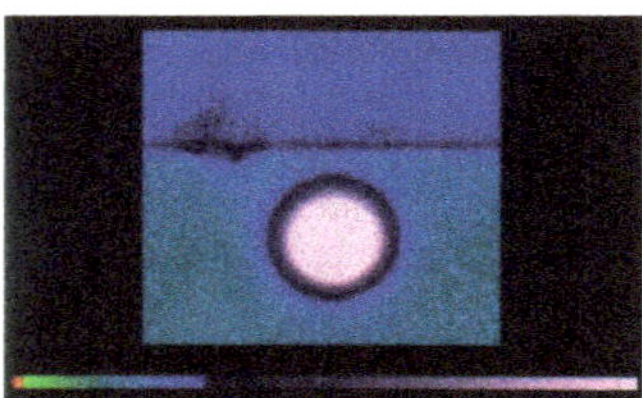

Figure 3. Image of a quasi-circular waveguide observed with an optical microscope, the glass is in light blue, the core is in pink, air is in dark blue.

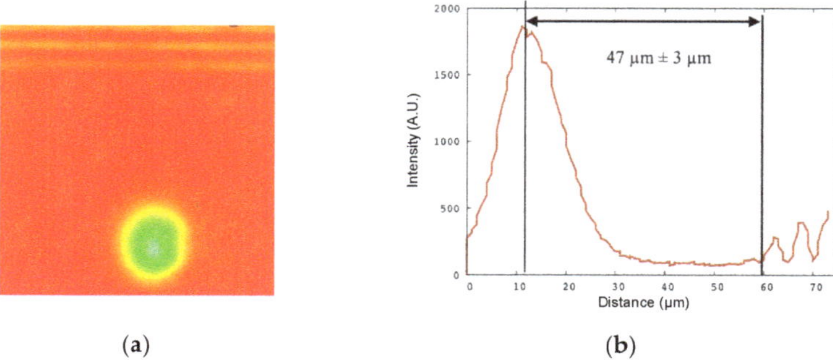

Figure 4. Realization of deeply buried waveguides by an Ag^+/Na^+ ion-exchange on a GO14 Teem-Photonics glass. The applied electric field during the burying process was 650 kV/m; (**a**) image of the output of the waveguide observed with an InGaAs Camera at $\lambda = 1.5$ µm; (**b**) vertical cut of the measured intensity showing the position of the mode with respect to the glass wafer substrate.

2.2. Modelling Ion-Exchanged Waveguides

Extensive work has been carried-out throughout the years to characterize and model ion-exchanges processes [18–22]. In this article, we will focus on a relatively simple description since it occurred to be reliable enough to allow us designing waveguides and predicting their optical behavior efficiently. Ion-exchange can be seen as a two-step process: first the exchange itself that occurs at the surface of the glass and creates a normalized concentration c_s of doping ions. For thin film sources, this concentration is linked to the applied current by the following relation:

$$\frac{\partial c_s}{\partial x} = \frac{J_0(c_s - 1)}{D_{Ag}},$$

(3)

where J_0 is the ion flux created by the electrolysis, x is the direction normal to the surface, and D_{Ag} is the diffusion coefficient of silver in the glass.

For liquid sources made of a mixture of molten salts containing B^+ and A^+ ions in order to replace A^+ ions of the glass, an equilibrium at the glass surface is usually rapidly reached, according to the chemical reaction:

$$A^+_{salt} + B^+_{glass} \Leftrightarrow B^+_{salt} + A^+_{glass},$$

(4)

Considering that the amount of ions in the molten salt is much bigger than the one of the glass, the ion concentrations in the liquid source can be considered as constant, which allows deriving the relative concentration at the surface:

$$c_s = \frac{K x_B}{1 + x_B(K - 1)},$$

(5)

K being the equilibrium constant of the chemical reaction (4) and $x_B = C_B^{salt} / (C_B^{salt} + C_A^{salt})$ is the molar fraction of doping ions B^+ in the molten salt.

Since the refractive index is proportional to the relative concentration, according to (2), it is easy to fix the refractive index change at the glass surface by setting the ratio of B^+ ions in the liquid source. Figure 5 shows an experimental determination of this dependence for a silver/sodium ion-exchange on a Schott-BF33 glass. These data have been obtained by realizing highly multimode slab waveguides and retrieving their refractive index profile through m-lines measurements [23] and the Inv-WKB procedure [24,25].

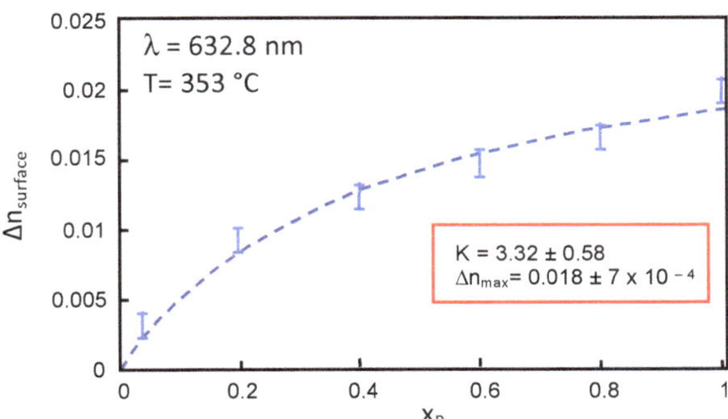

Figure 5. Refractive index change measured at the surface of a Schott-BF33 glass for different $x_B \text{AgNO}_3 + (1 - x_B)\text{NaNO}_3$ molten salts at a temperature of 353 °C.

The ions exchanged at the glass surface entail a gradient of concentration inside the glass. Hence, B^+ ions migrate inside the glass while A^+ ions are moving towards the surface. Since the two species of ions have different mobilities, an internal electrical field $\overrightarrow{E_{int}}$ is created during the diffusion process. To this field an external applied field $\overrightarrow{E_{app}}$ can be added, which results in ions fluxes $\overrightarrow{J_A}$ and $\overrightarrow{J_B}$, for A$^+$ and B$^+$, respectively, which are determined by the Nernst–Einstein equation:

$$\begin{aligned} \overrightarrow{J_A} &= -D_A\left[\overrightarrow{\nabla}C_A - \frac{e}{H\,k_BT}C_A\left(\overrightarrow{E_{int}} + \overrightarrow{E_{app}}\right)\right] \\ \overrightarrow{J_B} &= -D_B\left[\overrightarrow{\nabla}C_B - \frac{e}{H\,k_BT}C_B\left(\overrightarrow{E_{int}} + \overrightarrow{E_{app}}\right)\right] \end{aligned},$$

(6)

where D_i is the diffusion coefficient of the ion i, C_i its concentration, e is the electron charge, k_B the Boltzmann constant, T the temperature and H the Haven coefficient. Assuming that all the sites left by ions A$^+$ are filled by ions B$^+$, it can be written that at any position in the glass the relation $C_A + C_B = C_{A0}$, where C_{A0} is the concentration of A$^+$ ions before the exchange, is always valid. With this relation and Equation (6), the total ionic flux can be expressed as:

$$\overrightarrow{J} = \overrightarrow{J_A} + \overrightarrow{J_B} = -D_A C_{A0}\left[\alpha\overrightarrow{\nabla}c - \frac{e}{Hk_BT}(1 - \alpha c)\left(\overrightarrow{E_{int}} + \overrightarrow{E_{app}}\right)\right],$$

(7)

where the Steward coefficient $\alpha = 1 - D_B/D_A$ and the normalized concentration $c = C_B/C_{A0}$ have been introduced.

If no electric field is applied, then the total current is null, which allows determining easily $\overrightarrow{E_{int}}$:

$$\overrightarrow{E_{int}} = -\frac{Hk_BT}{e}\frac{\alpha\overrightarrow{\nabla}c}{1 - \alpha c}$$

(8)

The second Fick's law implies that:

$$\frac{\partial C_B}{\partial t} = -\nabla\overrightarrow{J_B}.$$

(9)

Combining (6), (8) and (9), the equation that governs the evolution of the relative concentration as a function of time is obtained:

$$\frac{\partial c}{\partial t} = \overrightarrow{\nabla}\left[\frac{D_B}{1 - \alpha c}\overrightarrow{\nabla}c - \frac{eD_B}{Hk_BT}c\overrightarrow{E_{app}}\right].$$

(10)

Equation (10) can be solved numerically by Finite Difference or Finite Element schemes but for accurate modelling, the dependence of ionic mobility and diffusion on the concentration should not be neglected. The so-called mixed alkali effect plays indeed a significant role in ion-exchanges where a high doping concentration is required [26,27]. It must also be noticed that ion-exchange modifies the conductivity of the glass, which in turn, modifies the field distribution of $\overrightarrow{E_{app}}$. Therefore, solving Equation (10) is actually much less obvious than it might appear and handling these problems has been the subject of a quite abundant literature [28–31]. Figure 6 displays typical refractive index profiles that have been obtained considering mixed alkali effect and the coupling between the ion-exchange and the applied electric field. Simulations have been done with an in-house software based on a finite difference scheme. It can be clearly seen how a proper choice of the experimental parameters can lead to circular waveguides. However, the maximum refractive index change is dropping from almost 0.1 to 10^{-2} during the burial process because of the spreading of doping ions caused by thermal diffusion.

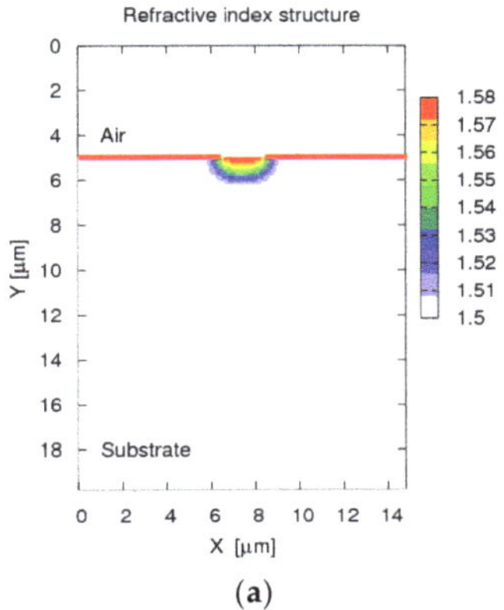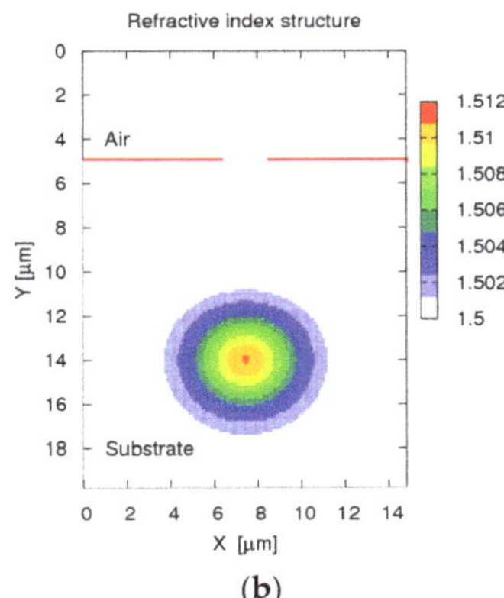

Figure 6. (**a**) Refractive index distribution of a thermally diffused waveguide, diffusion aperture width is 2 μm, exchange time is 2 min, $D_B = 0.8$ μm²/min; (**b**) refractive index profile of the waveguide (**a**) after an electrically assisted burying in a pure NaNO$_3$ molten salt, process duration is 1 h30 for an applied electric field of 180 kV/m.

2.3. Waveguide's Performances

The main characteristics when dealing with integrated optics waveguides are their spectral operation range, their losses that can be split between coupling and propagation losses, and their behavior with respect to light polarization.

2.3.1. Passive Glasses

Since the first waveguides demonstrated by Izawa and Nakagome [1], huge efforts have been made to reduce the losses of the waveguides. Historically, scattering represented the main source of losses. Indeed, the quality of the photolithography used for the realization of the masking layer before the ion-exchange was an issue as well as scratches or dirt deposited on the glass surface or refractive index inhomogeneities, such as bubbles. These problems are typical optical glass issues that are encountered when a custom-made glass is realized for the first time in small volumes, but they are easily handled by glass manufacturers when a higher volume of glass is produced. Therefore, state-of-the-art ion-exchanged waveguides are nowadays based on glass wafers specifically developed for this application or at least for microtechnologies. Among them, the more used are BF33 by Schott because of its compatibility with MEMS process, GO14 by TeemPhotonics SA and BGG31 by Schott [32], which have both been developed specifically for silver-sodium ion-exchanges. The interest of silver-sodium ion-exchange is that it allows the realization of buried waveguides solving, hence the problem of scattering due to surface defects or contaminations while dramatically improving the coupling efficiency with optical fibers. Nonetheless, silver-based technologies present also challenges since Ag$^+$ has a strong tendency to reduce into metallic Ag creating metallic clusters that are absorbing the optical signals. The glass composition should therefore be adapted not only to remove reducing elements like Fe, As, or Sb, but also to create a glass matrix where Na$^+$ ions are not linked to non-bridging oxygens [33]. The choice of the material for the masking layer should also be made with caution because the use of metallic mask can also induce the formation of Ag nanoparticles at the vicinity of the diffusion apertures [34]. Therefore, the use of Al or Ti mask is now often replaced by Al$_2$O$_3$ [35,36], SiO$_2$, or SiN [37] ones.

Table 1 presents the main characteristics of single mode waveguides realized on GO14, BGG31, and BF33, respectively. GO14 and BGG31 that have been optimized for telecom applications and ion-exchange present very low propagation losses and birefringence that are key characteristics for data transmission. BF33 is not a glass that has been designed for ion-exchange but it is a relatively low-cost glass that presents a quite good refractive index change and that is specifically indicated by its manufacturer for MEMS and microtechnology applications. Therefore, it is an excellent candidate for sensor realization and is mainly used for that. The relatively high propagation losses observed in BF33 is mainly due to the fact that this parameter is not very important in sensors and has, hence, neither been optimized nor measured accurately.

Table 1. Main characteristics of single mode waveguides realized on three different glasses.

Glass Type	GO14	BGG31	BF33
Losses	<0.05 dB/cm [7]	<0.1 dB/cm	<1 dB/cm (@780 nm)
Δn max	8×10^{-2} [7]	3.2×10^{-2} [38]	1.8×10^{-2} [35]
Birefringence	$<5 \times 10^{-4}$ [39]	$<2 \times 10^{-5}$ [40]	N.A.
Burying depth	~10 µm (50 µm max)	~10 µm [38]	~5 µm [35]

We deliberately did not mention Tl^+/K^+ ion-exchanged waveguides although the process is indeed the first one that has been used and the first one to be tentatively implemented in a production line. However, the advantages of a Tl^+/K^+ ion-exchange, namely a high refractive index change and the absence of clustering and absorption, are strongly counterbalanced by its toxicity, which implies dedicated safety procedures and waste treatments. It is therefore very scarcely used.

2.3.2. Active Glasses

The possibility of performing ion-exchange on rare-earth doped glasses was identified quite early. However, it was only in the 1990s with the development of WDM telecommunication that a lot of work was carried-out on the realization of efficient optical amplifiers and lasers. Because the solubility of rare earths into silicate glasses is quite low, which entails quenching due to clustering and reduces the amplifier efficiency, phosphate glasses rapidly emerged as the most efficient solution for obtaining high gain with compact devices. Among phosphate glasses, two specific references set the state of the art: they were the IOG 1 by Schott [41] and a proprietary glass referred as P1 by TeemPhotonics [42]. These two glasses succeeded in obtaining a high doping level without rare-earth clustering while being chemically resistant enough to withstand clean room processes and ion-exchange. The competition in the field of rare earth doped waveguides having been very hard, the characteristics of the different waveguides obtained in these glasses are difficult to find in the literature since the emphasis was mostly put on the active device performances, as will be detailed later.

2.3.3. Exotic Substrates

Some exotic glasses like fluoride glasses [43] or germanate glasses [44,45] have also been used for the realization of ion-exchanged waveguides but the difficulty in making sufficiently good wafers available at a reasonable cost, strongly limited the research in these directions.

3. Telecom Devices

3.1. Context and Historical Overview

Optical Telecommunications was originally the reason why Miller introduced the concept of integrated optics in 1969 [46]. Therefore, the pioneering work of integrated photonics on glass has been mainly devoted to telecommunication devices pushing steadily towards the development of not only ion-exchange processes but also of a full technology starting from the wafer fabrication and ending with the packaging of the manufactured Planar Lightwave Circuits. Figure 7 shows this evolution by displaying on one side one of the first demonstrations of a 1 to 8 power splitter made by cascading multimode Y-junctions [47] and, on the other side, its 2006 commercially available counterpart, single mode and Telcordia 1209 and 1221 compliant [7,48].

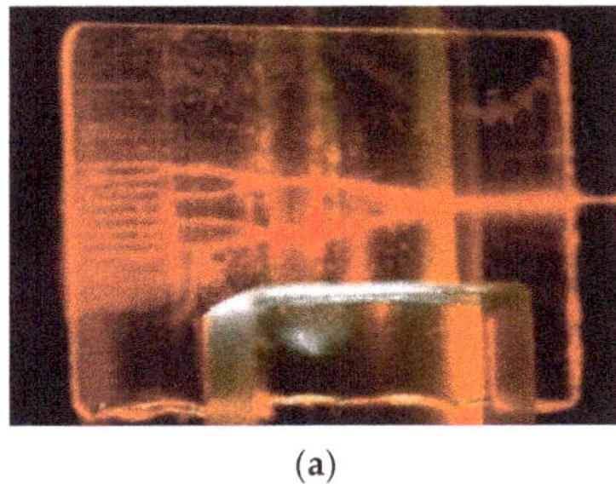 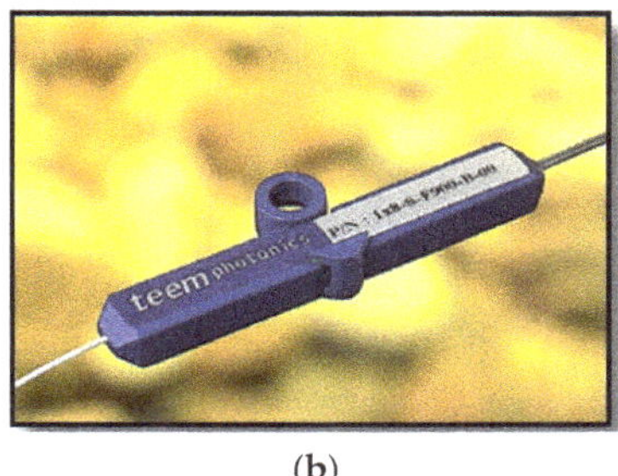

(**a**) (**b**)

Figure 7. 1 to 8 power splitter made by ion-exchange on glass (**a**) early demonstration in 1986; (**b**) qualified pigtailed and packaged commercially available product.

Once elementary functions, such as Y-junctions and directional couplers were demonstrated, studies were oriented towards all the functions that could be required for optical fiber communications like thermo-optic switches [49], Mach–Zehnder interferometers [50,51] and Multimode Mode Interference (MMI) couplers [52–55]. These buildings blocks have then been optimized and/or combined on a single chip to provide more functionality. In the next sections, we will review some of them and put the emphasis on the specificity brought by the use of ion-exchange on glass.

3.2. Wavelength Multiplexers

A five-channel wavelength demultiplexer-multiplexer has been demonstrated as early as 1982 by Suhara et al. using silver multimode waveguides combined with a Bragg grating [56]. More advanced devices using single mode waveguides include Arrayed-Waveguide Grating (AWG) multiplexers, whose quite large footprint is compensated by their low sensitivity to the light polarization thanks to the use of silver based buried waveguides [38]. A good thermal stability provided by the thickness of the glass substrate is also reported but a fine thermal tuning of the AWG's response remained possible [57]. Add and drop multiplexing has been achieved by combining Bragg gratings with Mach–Zehnder interferometers or more originally with a bimodal waveguide sandwiched by two asymmetric Y-branches [58]. Bragg grating can be integrated on glass by etching [59], wafer bonding [60], or photowriting [61–63].

Asymmetric Y-junctions are very interesting adiabatic devices that are well adapted to the smooth transitions between waveguides obtained by ion-exchange processes. Therefore, asymmetric Y-junctions have been used as stand-alone broadband wavelength multiplexers. For this type of applications, the asymmetry of the branches is obtained by a difference of the waveguide dimensions and a difference in their refractive index. Tailoring the refractive index of ion-exchanged waveguides can be achieved by segmenting the waveguide as demonstrated by Bucci et al. [64]. As can be seen on Figure 8, using vertical integration of deeply buried waveguides with selectively buried waveguides allowed obtaining a very broadband duplexing behavior while maintaining a relatively small surface footprint [36].

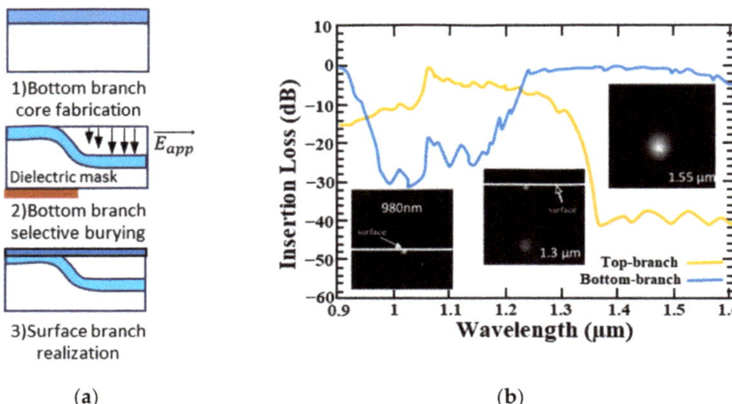

Figure 8. Vertically integrated broadband duplexer (**a**) fabrication steps; (**b**) measured transmission (the insets display the observed device output's mode at specific wavelengths). Top and bottom branches are separated by 28 µm [36].

3.3. Waveguide Amplifiers and Lasers

Active devices have been linked to the development of ion-exchanged devices since the beginning of this technology. Indeed, Saruwatari et al. demonstrated in 1973 a laser made with an optical amplifier based on a buried multimode ion-exchanged waveguide realized in a neodymium-doped borosilicate glass [65]. However, research on active devices really became a major field of research with a strong competition at the beginning of the 1990s when a lot of studies were carried-out. Work was first concentrated on Nd-doped amplifiers and lasers emitting at 1.06 µm since the four energy levels pumping scheme of this transition made it easier to achieve a net gain with the 800 nm pumping diodes available at the moment [66–70]. With the rise of Wavelength Division Multiplexing systems, optical amplifiers and sources operating in the C+L band (from 1525 nm to 1610 nm) became key devices and research on rare-earth doped integrated devices switched to the use of erbium ions whose transitions from the $^4I_{13/2}$ level to the $^4I_{15/2}$ one is broad enough to cover this wavelength range. Dealing with Er^{3+} active ions, the main issue was to realize waveguides with low-losses and a good overlap of the pump and signal modes. Indeed, the pumping scheme of this rare earth being a three levels one, the $^4I_{15/2}$ ground state absorbs the optical signal when it is not sufficiently pumped. Barbier et al. managed to solve this problem by developing a silver-sodium ion-exchange in their Er/Yb co-doped P1 glass [42]. 41 mm-long buried waveguides achieved 7 dB of net gain in a double pass configuration. This work has been followed by the demonstration of an amplifying four wavelength combiner [71] and the qualification of Erbium Doped Waveguide Amplifiers (EDWAs) in a 160 km-long WDM metro network [72]. This work has been completed by packaging and qualification developments in order to create a product line commercialized by TeemPhotonics.

Meanwhile the phosphate glasses developed by Schott also gained a lot of attention. Patel et al. achieved a record high gain of 13.7 dB/cm in a 3 mm-long waveguide realized by a silver film ion-exchange [73]. Such a gain per length unit was made possible by a high doping level of the glass in Er (8 wt. %) and Yb (12 wt. %).

Er-doped waveguide amplifiers being available, Er-doped lasers followed. Actually, the first proof of concept of an ion-exchanged waveguide laser was obtained on a modified BK7-silicate glass containing 0.5 wt. % of Er, with a potassium ion-exchange and two thin-film dielectric mirrors bonded to the waveguide's facets forming a Fabry-Perot cavity [74]. Nonetheless, from a strict point of view, this device was not a fully integrated laser because the mirrors were not integrated on the chip. Therefore, the next generation of Er-laser relied on the use of Bragg gratings as mirrors. In Distributed FeedBack (DFB) or Distribute Bragg

Reflectors configurations, these lasers presented a single frequency emission compatible with their use as transmitters in WDM systems. Similar for waveguide amplifiers, the use of phosphate glass entailed a major breakthrough in the performances. DBR lasers were demonstrated by Veasey et al. using a potassium ion-exchange [41], while Madasamy et al. manufactured similar devices with a silver thin film [75]. These approaches allowed integrating several lasers on a single chip to provide arrays of multiwavelength sources with one single grating, the wavelength selection being made by tuning the effective indices of the waveguides through their dimensions. Thanks to the use of highly concentrated molten salt of silver nitrate and a DFB configuration, Blaize et al. succeeded in creating a comb of 15 lasers with one single Bragg grating [76]. The emitters' wavelengths were spaced by 25 GHz and 100 GHz and set to be on the Dense WDM International Telecommunication Union (ITU) grid. The output power of these devices could be as high as 80 mW for a 350 mW coupled pump power [41], while a linewidth of only 3 kHz has been reported by Bastard et al. on their DFB lasers [77]. Figure 9 displays a picture of such a DFB laser pigtailed to HI1060 single mode fibers. The stability and purity of the emission of erbium doped waveguide lasers has been recently used to generate a Radio Frequency signal and successfully transmit data at a frequency of 60 GHz [78].

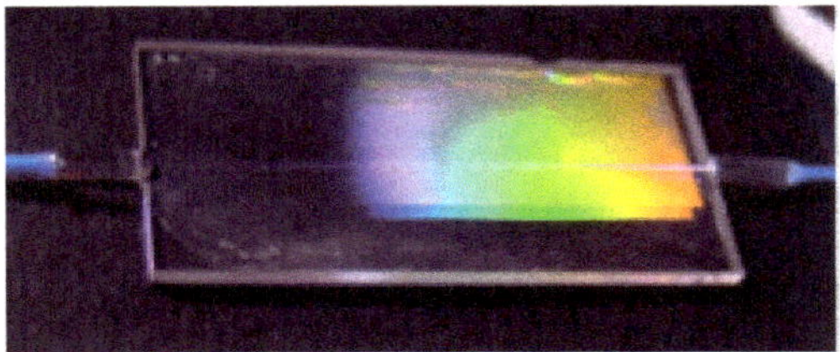

Figure 9. Picture of a DFB laser realized by silver-sodium ion-exchange on P1 phosphate glass at the IMEP-LaHC (device similar to [77]).

Bragg gratings on phosphate glass can be made by photolithography steps and etching like in [41,76,77] or by direct UV inscription like in [79,80] and on IOG1. The use of a hybrid un-doped/doped IOG1 substrate allowed Yliniemi et al. [80] to realize UV-written Bragg gratings with high reflectance and selectivity, demonstrating hence a single frequency emission with an output power of 9 mW and a slope efficiency of 13.9%.

3.4. Hybrid Devices

Ion-exchanged waveguides being made inside the glass wafer, they leave its surface plane and available for the integration of other materials or technologies. The realization of deeply buried waveguides [81] and selectively buried waveguides [82] acting as optical vias between two different layers increased furthermore the possibility of 3D integration. In order to overcome the quite weak chemical durability of an Yb-Er doped phosphate glass, Gardillou et al. [83] wafer bonded it on a silicate glass substrate containing surface Tl ion-exchanged strips. The higher refractive index active glass was then thinned by an appropriate polishing process to become a single mode planar waveguide. At the place where the planar waveguide was in contact with the ion-exchanged strips, the variation of refractive index provided the lateral confinement creating hence a hybrid waveguide. A gain of 4.25 dB/cm has been measured with this device. This approach has been pursued by Casale et al. [59] who realized a hybrid DFB laser combining a planar ion-exchanged waveguide made on IOG1 with a passive ion-exchanged channel waveguide realized on GO14. The Bragg grating was etched on the passive glass and encapsulated between the two wafers.

Polymers have also been used to functionalize an ion-exchanged waveguide. As an example, a thin film of BDN-doped cellulose acetate deposited on the surface of ion-exchanged waveguide lasers allowed the realization of passively Q-switched lasers on

Nd-doped [84] and Yb doped [85] IOG1 substrates. A peak power of 1 kW for pulses of 1.3 ns and a repetition rate of 28 kHz has been reported by Charlet et al. [86] and used successfully to pump a photonic crystal fiber and generate a supercontinuum [87].

Recently, a proof of concept of LiNbO$_3$ thin films hybridized on ion-exchanged waveguides have been reported [88]. The combination of these two well-known technological platforms for integrated photonics opens the route towards efficient low-loss non-linear integrated devices including electro-optic modulators.

Hybrid integration of semiconductor devices on glass wafers containing ion-exchanged waveguides have been reported for the first time in 1987, by MacDonald et al. [89] They bonded GaAs photodiodes on a metallic layer previously deposited and patterned on the glass wafer. Waveguides were done by a silver thin film dry process. Silicon [90] and germanium [91] photodetectors have been produced on potassium waveguides, while Yi-Yan et al. proposed a lift-off approach to bound thin III-V semiconductor membranes on the surface of a glass wafer containing ion-exchanged waveguide and realize Metal–Semiconductor–Metal (MSM) photodetectors [92].

4. Sensors

Integrated photonics is intrinsically interesting for the realization of optical sensors because it provides compact and reliable self-aligned devices that can be easily deported when pigtailed to optical fibers. Glass is a material that is chemically inert, bio-compatible, and mechanically stable. Therefore, making optical sensors on glass wafers or integrating optical glass chips into complex set-ups have encountered a huge interest. We will detail here a selection of ion-exchanged based glass sensors as examples of possible applications.

Although AWGs used in telecom are actually integrated spectrometers, they are not well adapted to the rapid measurement of full spectra. For this reason, a Stationary-Wave Integrated Fourier-Transform Spectrometer (SWIFTS) has been proposed and developed [93]. It is a static Fourier Spectrometer that measures directly the intensity of a standing wave with nanoprobe placed on a waveguide. In the instrument reported by Thomas et al. [94], the waveguide is made by a silver ion-exchange on a silicate glass and the nanoprobes are gold nanodots. The interaction of gold nano-antennas with an ion-exchanged waveguide has been studied by Arnaud et al. [95]. This spectrometer has a spectral measurement range that starts at 630 nm and ends at 1080 nm with a spectral resolution better than 14 pm. SWIFTS interferometers are currently integrated in the product line commercialized by Resolution Spectra Systems [96].

Displacement sensors allow measuring accurately the change of position of an object through interferometry. Helleso et al. [97] implemented a double Michelson interferometer on a glass substrate using potassium ion-exchange; the device provided two de-phased outputs in order to give access not only to the distance of the displacement but also its direction. However, having only two interferometric signals is not sufficient to prevent the measure from being affected by unexpected signal variations. For this reason, Lang et al. [98] proposed a new design for the interferometric head that provided four quadrature phase shifted outputs. The device made by potassium ion-exchange demonstrated a measurement accuracy of 79 nm over a measurement range of several meters when used with an HeNe laser as a source. After technological improvements and the use of a silver-sodium ion-exchange on GO14 glass, an evolution of this sensor is now commercialized by TeemPhotonics and presents a resolution of 10 pm for a 1530 nm–1560 nm operating wavelength range [48].

Measuring speed is also something that can be of major importance, specifically in the case of aircrafts where their True Air Speed (TAS), which is their speed with respect to the air surrounding them, conditions their lift. Airborne LIDARs have hence been developed as a backup to Pitot gauges in order to increase the safety of flight by providing a redundant accurate measurement of the aircraft TAS. The operation principle of an airborne LIDAR is based on the Doppler frequency shift measured on a laser signal reflected on the dust particles of the atmosphere. This shift being quite low and presenting a low amplitude when compared to the emitted signal, a laser source that presents a narrow linewidth, a low Relative Intensity Noise and that is resilient to mechanical vibrations is required. Bastard et al. [99] realized such a laser source on an Er/Yb doped phosphate glass with silver ion-exchanged waveguides and a DFB structure. This laser presented a fiber coupled output power of 2.5 mW, a linewidth of 2.5 kHz, and a RIN that was 6 dB lower than the specification limit. The device has then been successfully implemented in the LIDAR set-up and validated in flight [100].

Astrophysical research programs rely on telescopes with always higher resolution to detect exoplanets, young star accretion disks, etc. Optical long baseline instruments, which interferometrically combine the signal collected by different telescope have been developed for this purpose. Such complex interferometers are very sensitive to misalignment and vibrations, therefore the use of integrated optics as telescope recombiners have been studied. Haguenauer et al. [101] used a silver-sodium ion-exchange on a silicate glass to realize a two telescope beam combiner operating on the H atmospheric band (from $\lambda = 1.43$ µm to $\lambda = 1.77$ µm). Consisting of a proper arrangement of three Y-junctions, the device had two photometric and one interferometric outputs. The fringe contrast obtained in the laboratory was 92% and the device was included in the Integrated Optic Near infrared Interferometric Camera (IONIC) put into a cryostat and successfully qualified on the sky [102]. Figure 10 shows the MAFL chip [103] that was developed for the interferometric combination of three telescopes. The pigtailed instrument contained not only the science interferometers but also three other ones dedicated to metrology, which permitted measuring of the different optical paths. The functions multiplexing and demultiplexing the metrology signal and the science ones were also implemented on the chip.

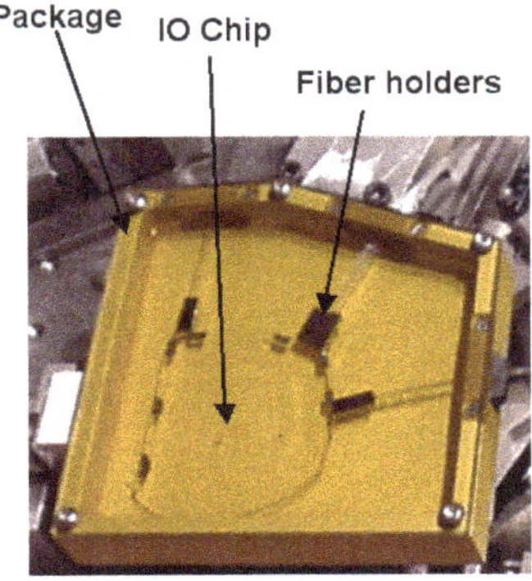

Figure 10. Picture of the MAFL combining module. The optical chip contains waveguides made by a silver sodium ion-exchange.

The chemical durability of silicate glasses is a major advantage when a use in harsh environment is required. The opto-fluidic sensor developed by Allenet et al. [104] represents a quite extreme example of this. Indeed, the ion-exchange technology developed by Schimpf et al. [35] on BF33 glass has been employed to realize a sensor for the detection of plutonium in a nuclear plant environment. The fully pigtailed and packaged device that is depicted on Figure 11, has been successfully tested in a nuclearized glove box, detecting plutonium dissolved in 2 Mol nitric acid without a failure over a period of one month. Such a reliability was achieved by co-integrating microfluidic channels fabricated by HF wet etching on one BF33 wafer with silver ion-exchanged waveguides realized on another wafer. The two wafers have been assembled by molecular adherence avoiding hence the use of radiation sensitive epoxy glues.

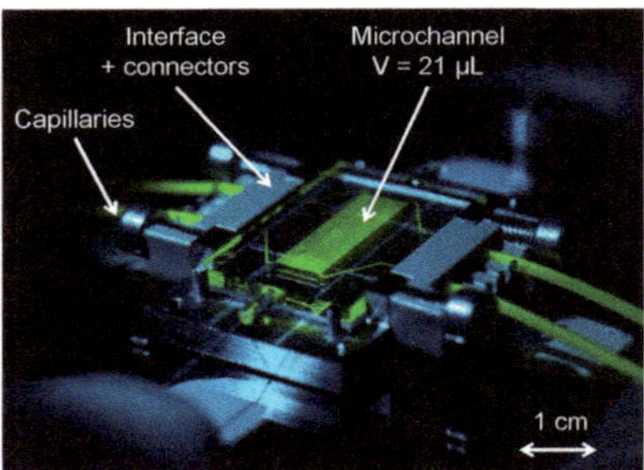

Figure 11. Picture of an optofluidic sensor realized on BF33 glass wafer for the measurement of radioactive elements diluted in highly concentrated nitric acid.

5. Conclusions

In this paper, we reviewed over thirty years of activities in glass photonics. The ion-exchange realization process as well as its modelling has been exposed. Passive and active devices for telecommunication applications have then been presented with the emphasis on the major breakthroughs of this field. The section dedicated to sensors underlines the evolution of the ion-exchange technology, which is moving from quite simple, though extremely performant functions, to more complex integrated optical microsystems. The authors hope that the picture of glass photonics that they presented will soon be outdated by the new results that are currently being elaborated in the many laboratories of universities and companies involved in this field throughout the world.

Funding: The visit of Pr Broquin at the University of Eastern Finland is funded by the Nokia Foundation within the frame of the Institut Français de Finlande—Nokia Foundation Distinguished Chair. This work is also part of the Academy of Finland Flagship Programme, Photonics Research and Innovation (PREIN), decision 320166.

Institutional Review Board Statement: Not applicable.

Informed Consent Statement: Not applicable.

Conflicts of Interest: The authors declare no conflict of interest.

References

1. Izawa, T. Optical waveguide formed by electrically induced migration of ions in glass plates. *Appl. Phys. Lett.* **1972**, *21*, 584–586. [CrossRef]
2. Terai, R.; Hayami, R. Ionic diffusion in glasses. *J. Non-Crystalline Solids* **1975**, *18*, 217–264. [CrossRef]
3. Findakly, T. Glass Waveguides by Ion Exchange: A Review. *Opt. Eng.* **1985**, *24*, 242244. [CrossRef]
4. Ramaswamy, R.; Srivastava, R. Ion-exchanged glass waveguides: A review. *J. Light. Technol.* **1988**, *6*, 984–1000. [CrossRef]
5. Ross, L. Integrated Optical Components in Substrate Glasses. *Glastech. Ber.* **1989**, *62*, 285–297.
6. Nikonorov, G.T.; Petrovskii, N.V. Ion-exchanged glasses in integrated optics: The current state of research and prospects (a review). *Glass Phys. Chem.* **1999**, *25*, 16–55.
7. Broquin, J.E. Glass integrated optics: State of the art and position toward other technologies. In Proceedings of the Integrated Optics: Devices, Materials, and Technologies XI, San Jose, CA, USA, 22–24 January 2007.
8. Mazzoldi, P.; Sada, C. A trip in the history and evolution of ion-exchange process. *Mater. Sci. Eng. B* **2008**, *149*, 112–117. [CrossRef]
9. West, B.R.; Tervonen, A.; Honkanen, S. Ion-exchanged glass waveguide technology: A review. *Opt. Eng.* **2011**, *50*, 071107. [CrossRef]
10. Zachariasen, W.H. The atomic arrangement in glass. *J. Am. Chem. Soc.* **1932**, *54*, 3841–3851. [CrossRef]
11. Huggins, M.L.; Sun, K.H.; Davis, D.O. The Dispersion of Silicate Glasses as a Function of Composition II*. *J. Opt. Soc. Am.* **1942**, *32*, 635–648. [CrossRef]
12. Doremus, R.H. *Glass Science*; Wiley & Sons: New York, NY, USA, 1973.
13. Lilienhof, H.J.; Voges, E.; Ritter, D.; Pantschew, B. Field-induced index profiles of multimode ion-exchanged strip waveguides. *IEEE J. Quantum Electron.* **1982**, *18*, 1877–1883. [CrossRef]
14. Chartier, G.; Jaussaud, P.; De Oliveira, A.; Parriaux, O. Fast fabrication method for thick and highly multimode optical waveguides. *Electron. Lett.* **1977**, *13*, 763–764. [CrossRef]

15. Viljanen, J.; Leppihalme, M. Fabrication of optical strip waveguides with nearly circular cross section by silver ion migration technique. *J. Appl. Phys.* **1980**, *51*, 3563–3565. [CrossRef]

16. Pantchev, B. One-step field-assisted ion exchange for fabrication of buried multimode optical strip waveguides. *Electron. Lett.* **1987**, *23*, 1188–1190. [CrossRef]

17. Seki, M.; Hashizume, H.; Sugawara, R. Two-step purely thermal ion-exchange technique for single-mode waveguide devices in glass. *Electron. Lett.* **1988**, *24*, 1258–1259. [CrossRef]

18. Prieto-Blanco, X. Electro-diffusion equations of monovalent cations in glass under charge neutrality approximation for optical waveguide fabrication. *Opt. Mater.* **2008**, *31*, 418–428. [CrossRef]

19. Quaranta, A.; Cattaruzza, E.; Gonella, F. Modelling the ion exchange process in glass: Phenomenological approaches and perspectives. *Mater. Sci. Eng. B* **2008**, *149*, 133–139. [CrossRef]

20. Quaranta, A.; Rahman, A.; Mariotto, G.; Maurizio, C.; Trave, E.; Gonella, F.; Cattaruzza, E.; Gibaudo, E.; Broquin, J.E. Spectroscopic Investigation of Structural Rearrangements in Silver Ion-Exchanged Silicate Glasses. *J. Phys. Chem. C* **2012**, *116*, 3757–3764. [CrossRef]

21. Albert, J.; Lit, J.W. Full modeling of field-assisted ion exchange for graded index buried channel optical waveguides. *Appl. Opt.* **1990**, *29*, 2798–2804. [CrossRef] [PubMed]

22. Tervonen, A. A general model for fabrication processes of channel waveguides by ion exchange. *J. Appl. Phys.* **1990**, *67*, 2746–2752. [CrossRef]

23. Tien, P.K.; Ulrich, R. Theory of Prism–Film Coupler and Thin-Film Light Guides. *J. Opt. Soc. Am.* **1970**, *60*, 1325–1337. [CrossRef]

24. Chiang, K. Construction of refractive-index profiles of planar dielectric waveguides from the distribution of effective indexes. *J. Light. Technol.* **1985**, *3*, 385–391. [CrossRef]

25. White, J.M.; Heidrich, P.F. Optical waveguide refractive index profiles determined from measurement of mode indices: A simple analysis. *Appl. Opt.* **1976**, *15*, 151–155. [CrossRef] [PubMed]

26. Kirchheim, R.; Paulmann, D. The relevance of site energy distribution for the mixed alkali effect. *J. Non-Cryst. Solids* **2001**, *286*, 210–223. [CrossRef]

27. Lupascu, A.; Kevorkian, A.; Boudet, T.; Saint-Andre, F.; Persegol, D.; Levy, M. Modeling ion exchange in glass with concentration-dependent diffusion coefficients and mobilities. *Opt. Eng.* **1996**, *35*, 1603–1610. [CrossRef]

28. West, B.R.; Madasamy, P.; Peyghambarian, N.; Honkanen, S. Modeling of ion-exchanged glass waveguide structures. *J. Non-Cryst. Solids* **2004**, *347*, 18–26. [CrossRef]

29. Hazart, J.; Minier, V. Concentration profile calculation for buried ion-exchanged channel waveguides in glass using explicit space-charge analysis. *IEEE J. Quantum Electron.* **2001**, *37*, 606–612. [CrossRef]

30. Masalkar, P.J. Calculation of Concentration Profile in Ion-Exchange Waveguides by Finite Difference ADI Method. *Optik* **1994**, *95*, 168–172.

31. Saarikoski, H.; Salmio, R.P.; Saarinen, J.; Eirola, T.; Tervonen, A. Fast numerical solution of nonlinear diffusion equation for the simulation of ion-exchanged micro-optics components in glass. *Opt. Commun.* **1997**, *134*, 362–370. [CrossRef]

32. Fabricius, N.; Oeste, H.; Guttmann, H.; Quast, H.; Ross, L. BGG 31: A New Glass for Multimode Waveguide Fabrication. *Proc. EFOC/LAN* **1988**, *88*, 59–62.

33. Araujo, R. Colorless glasses containing ion-exchanged silver. *Appl. Opt.* **1992**, *31*, 5221–5224. [CrossRef] [PubMed]

34. Walker, R.G.; Wilkinson, C.D.W.; Wilkinson, J.A.H. Integrated optical waveguiding structures made by silver ion-exchange in glass 1: The propagation characteristics of stripe ion-exchanged waveguides: A theoretical and experimental investigation. *Appl. Opt.* **1983**, *22*, 1923–1928. [CrossRef] [PubMed]

35. Schimpf, A.; Bucci, D.; Nannini, M.; Magnaldo, A.; Couston, L.; Broquin, J.-E. Photothermal microfluidic sensor based on an integrated Young interferometer made by ion exchange in glass. *Sens. Actuators B Chem.* **2012**, *163*, 29–37. [CrossRef]

36. Onestas, L.; Bucci, D.; Ghibaudo, E.; Broquin, J.-E. Vertically Integrated Broadband Duplexer for Erbium-Doped Waveguide Amplifiers Made by Ion Exchange on Glass. *IEEE Photonics Technol. Lett.* **2011**, *23*, 648–650. [CrossRef]

37. Pantchev, B.; Danesh, P. Masking Problem in the Fabrication of Optical Waveguide Structures in Glass by Double Ion Exchange. *Jpn. J. Appl. Phys.* **1997**, *36*, 4320–4322. [CrossRef]

38. Glingener, B.B.C. Polarization Insensitive Ion-Exchanged Arrayed-Waveguide Grating Multiplexers in Glass. *Fiber Integr. Opt.* **1998**, *17*, 279–298. [CrossRef]

39. Jamon, D.; Royer, F.; Parsy, F.; Ghibaudo, E.; Broquin, J.E. Birefringence Measurements in Optical Waveguides. *J. Light. Technol.* **2013**, *31*, 3151–3157. [CrossRef]

40. Yliniemi, S.; West, B.R.; Honkanen, S. Ion-exchanged glass waveguides with low birefringence for a broad range of waveguide widths. *Appl. Opt.* **2005**, *44*, 3358–3363. [CrossRef] [PubMed]

41. Veasey, D.L.; Funk, D.S.; Sanford, N.A.; Hayden, J.S. Arrays of distributed-Bragg-reflector waveguide lasers at 1536 nm in Yb/Er codoped phosphate glass. *Appl. Phys. Lett.* **1999**, *74*, 789–791. [CrossRef]

42. Barbier, D.; Delavaux, J.M.; Kevorkian, A.; Gastaldo, P.; Jouanno, J.M. Yb/Er Integrated optics amplifiers on phosphate glass in single and double pass configurations. In Proceedings of the Optical Fiber Communication Conference, San Diego, CA, USA, 26 February 1995.

43. Josse, E.; Broquin, J.E.; Lebrasseur, E.; Fonteneau, G.; Rimet, R.; Jacquier, B.; Lucas, J. *Rare-Earth-Doped Fluoride Waveguides*; International Society for Optics and Photonics: Bellingham, WA, USA, 1997; Volume 2996, pp. 74–85.

44. Grelin, J.; Bouchard, A.; Ghibaudo, E.; Broquin, J.-E. Study of Ag+/Na+ ion-exchange diffusion on germanate glasses: Realization of single-mode waveguides at the wavelength of 1.55 μm. *Mater. Sci. Eng. B* **2008**, *149*, 190–194. [CrossRef]
45. Luo, T.; Jiang, S.; Conti, G.N.; Honkanen, S.; Mendes, S.; Peyghambarian, N. Ag/Sup+/-Na/Sup+/Exchanged Channel Waveguides in Germanate Glass. *Electron. Lett.* **1998**, *34*, 2239–2240. [CrossRef]
46. Miller, S.E. Integrated Optics: An Introduction. *Bell Syst. Tech. J.* **1969**, *48*, 2059–2069. [CrossRef]
47. Voirin, G.; Rimet, R.; Chartier, G. Performances of an Ion Exchanged Star Coupler for Multimode Optical Communications. In *Optical Sciences*; Springer: Berlin/Heidelberg, Germany, 1985; Volume 48, pp. 229–231.
48. Teem Photonics. PIC Optical Waveguides for Si-Pho, Integration & Sensors. Available online: https://www.teemphotonics.com/integrated-photonics/platform/ (accessed on 25 April 2021).
49. Hurvitz, T.; Ruschin, S.; Brooks, D.; Hurvitz, G.; Arad, E. Variable optical attenuator based on ion-exchange technology in glass. *J. Light. Technol.* **2005**, *23*, 1918–1922. [CrossRef]
50. Tervonen, A.; Honkanen, S.; Najafi, S.I. Analysis of symmetric directional couplers and asymmetric Mach-Zehnder interferometers as 1.30- and 1.55-um dual-wavelength demultiplexers/multiplexers. *Opt. Eng.* **1993**, *32*, 2083–2091. [CrossRef]
51. Benech, P.; Persegol, D.; Andre, F.S. A glass ion exchanged Mach-Zehnder interferometer to stabilise the frequency of a laser diode. *J. Phys. D Appl. Phys.* **1990**, *23*, 617–619. [CrossRef]
52. Das, S.; Geraghty, D.F.; Peyghambarian, N. MMI splitters by ion-exchange in glass. In *Proceedings of the Integrated Optics Devices IV*; International Society for Optics and Photonics: Bellingham, WA, USA, 2000; Volume 3936, pp. 239–248.
53. Kasprzak, D.; Błahut, M.; Maciak, E. Applications of multimode interference effects in gradient waveguides produced by ion-exchange in glass. *Eur. Phys. J. Spéc. Top.* **2008**, *154*, 113–116. [CrossRef]
54. West, B.R.; Honkanen, S. MMI devices with weak guiding designed in three dimensions using a genetic algorithm. *Opt. Express* **2004**, *12*, 2716–2722. [CrossRef] [PubMed]
55. West, B.R.; Plant, D.V. Optimization of non-ideal multimode interference devices. *Opt. Commun.* **2007**, *279*, 72–78. [CrossRef]
56. Suhara, T.; Viljanen, J.; Leppihalme, M. Integrated-optic wavelength multi- and demultiplexers using a chirped grating and an ion-exchanged waveguide. *Appl. Opt.* **1982**, *21*, 2195–2198. [CrossRef] [PubMed]
57. Ruschin, S.; Hurwitz, G.; Hurwitz, T.; Kepten, A.; Arad, E.; Soreq, Y.; Eckhouse, S. Glass ion-exchange technology for wavelength management applications. *Photonics Fabr. Eur.* **2003**, *4944*, 150–159. [CrossRef]
58. Castro, J.M.; Geraghty, D.F.; West, B.R.; Honkanen, S. Fabrication and comprehensive modeling of ion-exchanged Bragg optical add-drop multiplexers. *Appl. Opt.* **2004**, *43*, 6166–6173. [CrossRef] [PubMed]
59. Casale, M.; Bucci, D.; Bastard, L.; Broquin, J.-E.; Quaranta, A. Hybrid erbium-doped DFB waveguide laser made by wafer bonding of two ion-exchanged glasses. *Ceram. Int.* **2015**, *41*, 7466–7470. [CrossRef]
60. Gardillou, F.; Bastard, L.; Broquin, J.E. Integrated optics Bragg filters made by ion exchange and wafer bonding. *Appl. Phys. Lett.* **2006**, *89*, 101123. [CrossRef]
61. Román, J.E.; Winick, K.A. Photowritten gratings in ion-exchanged glass waveguides. *Opt. Lett.* **1993**, *18*, 808–810. [CrossRef] [PubMed]
62. Montero, C.; Gomez-Reino, C.; Brebner, J.L. Planar Bragg gratings made by excimer-laser modification of ion-exchanged waveguides. *Opt. Lett.* **1999**, *24*, 1487–1489. [CrossRef] [PubMed]
63. Geraghty, D.; Provenzano, D.; Marshall, W.; Honkanen, S.; Yariv, A.; Peyghambarian, N. Gratings photowritten in ion-exchanged glass channel waveguides. *Electron. Lett.* **1999**, *35*, 585. [CrossRef]
64. Bucci, D.; Grelin, J.; Ghibaudo, E.; Broquin, J.-E. Realization of a 980-nm/1550-nm Pump-Signal (De)multiplexer Made by Ion-Exchange on Glass Using a Segmented Asymmetric Y-Junction. *IEEE Photonics Technol. Lett.* **2007**, *19*, 698–700. [CrossRef]
65. Saruwatari, M. Nd-glass laser with three-dimensional optical waveguide. *Appl. Phys. Lett.* **1974**, *24*, 603–605. [CrossRef]
66. Aoki, H.; Ishikawa, E.; Asahara, Y. Nd/Sup 3+/-Doped Glass Waveguide Amplifier at 1.054 Mu m. *Electron. Lett.* **1991**, *27*, 2351–2353. [CrossRef]
67. Sanford, N.A.; Malone, K.J.; Larson, D.R.; Hickernell, R.K. Y-branch waveguide glass laser and amplifier. *Opt. Lett.* **1991**, *16*, 1168–1170. [CrossRef] [PubMed]
68. Miliou, A.; Cao, X.; Srivastava, R.; Ramaswamy, R. 15-dB amplification at 1.06 mu m in ion-exchanged silicate glass waveguides. *IEEE Photonics Technol. Lett.* **1993**, *5*, 416–418. [CrossRef]
69. Sanford, N.A.; Aust, J.A.; Malone, K.J.; Larson, D.R. Linewidth narrowing in an imbalanced Y-branch waveguide laser. *Opt. Lett.* **1993**, *18*, 281–283. [CrossRef] [PubMed]
70. Aust, J.A.; Malone, K.J.; Veasey, D.L.; Sanford, N.A.; Roshko, A. Passively Q-switched Nd-doped waveguide laser. *Opt. Lett.* **1994**, *19*, 1849–1851. [CrossRef]
71. Barbier, D.; Rattay, M.; Andre, F.S.; Clauss, G.; Trouillon, M.; Kevorkian, A.; Delavaux, J.M.; Murphy, E. Amplifying four-wavelength combiner, based on erbium/ytterbium-doped waveguide amplifiers and integrated splitters. *IEEE Photonics Technol. Lett.* **1997**, *9*, 315–317. [CrossRef]
72. Reichmann, K.; Iannone, P.; Birk, M.; Frigo, N.; Barbier, D.; Cassagnettes, C.; Garret, T.; Verlucco, A.; Perrier, S.; Philipsen, J. An eight-wavelength 160-km transparent metro WDM ring network featuring cascaded erbium-doped waveguide amplifiers. *IEEE Photonics Technol. Lett.* **2001**, *13*, 1130–1132. [CrossRef]
73. Patel, F.; Dicarolis, S.; Lum, P.; Venkatesh, S.; Miller, J. A Compact High-Performance Optical Waveguide Amplifier. *IEEE Photonics Technol. Lett.* **2004**, *16*, 2607–2609. [CrossRef]

74. Feuchter, T.; Mwarania, E.; Wang, J.; Reekie, L.; Wilkinson, J. Erbium-doped ion-exchanged waveguide lasers in BK-7 glass. *IEEE Photonics Technol. Lett.* **1992**, *4*, 542–544. [CrossRef]
75. Madasamy, P.; Conti, G.N.; Poyhonen, P.; Hu, Y.; Morrell, M.M.; Geraghty, D.F.; Honkanen, S.; Peyghambarian, N. Wave-guide Distributed Bragg Reflector Laser Arrays in Erbium Doped Glass Made by Dry Ag Film Ion Exchange. *Opt. Eng.* **2002**, *41*, 1084–1086.
76. Blaize, S.; Bastard, L.; Cassagnetes, C.; Broquin, J. Multiwavelengths DFB waveguide laser arrays in Yb-Er codoped phosphate glass substrate. *IEEE Photonics Technol. Lett.* **2003**, *15*, 516–518. [CrossRef]
77. Bastard, L.; Blaize, S.; Broquin, J.-E. Glass integrated optics ultranarrow linewidth distributed feedback laser matrix for dense wavelength division multiplexing applications. *Opt. Eng.* **2003**, *42*, 2800–2804. [CrossRef]
78. Arab, N.; Bastard, L.; Poëtte, J.; Broquin, J.-E.; Cabon, B. Thermal coupling impact on an MMW carrier generated using two free-running DFB lasers on glass. *Opt. Lett.* **2018**, *43*, 5500–5503. [CrossRef] [PubMed]
79. Yliniemi, S.; Honkanen, S.; Laronche, A.; Albert, J.; Ianoul, A. Photosensitivity and volume gratings in phosphate glasses for rare-earth-doped ion-exchanged optical waveguide lasers. *J. Opt. Soc. Am. B* **2006**, *23*, 2470–2478. [CrossRef]
80. Yliniemi, S.; Albert, J.; Wang, Q.; Honkanen, S. UV-exposed Bragg gratings for laser applications in silver-sodium ion-exchanged phosphate glass waveguides. *Opt. Express* **2006**, *14*, 2898–2903. [CrossRef] [PubMed]
81. Grelin, J.; Ghibaudo, E.; Broquin, J.-E. Study of deeply buried waveguides: A way towards 3D integration. *Mater. Sci. Eng. B* **2008**, *149*, 185–189. [CrossRef]
82. Bertoldi, O.; Broquin, J.E.; Vitrant, G.; Collomb, V.; Trouillon, M.; Minier, V. Use of Selectively Buried Ion-Exchange Wave-guides for the Realization of Bragg Grating Filters. In *Proceedings of the Integrated Optics and Photonic Integrated Circuits*; International Society for Optics and Photonics: Bellingham, WA, USA, 2004; Volume 5451, pp. 182–190.
83. Gardillou, F.; Bastard, L.; Broquin, J.-E. 4.25 dB gain in a hybrid silicate/phosphate glasses optical amplifier made by wafer bonding and ion-exchange techniques. *Appl. Phys. Lett.* **2004**, *85*, 5176. [CrossRef]
84. Salas-Montiel, R.; Bastard, L.; Grosa, G.; Broquin, J.-E. Hybrid Neodymium-doped passively Q-switched waveguide laser. *Mater. Sci. Eng. B* **2008**, *149*, 181–184. [CrossRef]
85. Ouslimani, H.; Bastard, L.; Broquin, J.-E. Narrow-linewidth Q-switched DBR laser on Ytterbium-doped glass. *Ceram. Int.* **2015**, *41*, 8650–8654. [CrossRef]
86. Charlet, B.; Bastard, L.; Broquin, J.E. 1 kW peak power passively Q-switched Nd(3+)-doped glass integrated waveguide laser. *Opt. Lett.* **2011**, *36*, 1987–1989. [CrossRef]
87. Charlet, B.; Bastard, L.; Broquin, J.E.; Bucci, D.; Ghibaudo, E. Supercontinuum Source Based on Ion-Exchanged Neodymium Doped Q-Switched Laser. In Proceedings of the European Conference on Integrated Optics, Sitges, Spain, 18–20 April 2012.
88. Legrand, L.; Bouchard, A.; Grosa, G.; Broquin, J.E. Hybrid integration of 300nm-thick LiNbO3 films on ion-exchanged glass waveguides for efficient nonlinear integrated devices. In *Proceedings of the Integrated Optics: Devices, Materials, and Technologies XXII*; International Society for Optics and Photonics: Bellingham, WA, USA, 2018; Volume 10535, p. 105350E.
89. Macdonald, R.I.; Lam, D.K.W.; Syrett, B.A. Hybrid optoelectronic integrated circuit. *Appl. Opt.* **1987**, *26*, 842–844. [CrossRef] [PubMed]
90. Larson, D.R.; Phelan, R.J., Jr. Fast Optical Detector Deposited on Dielectric Channel Waveguides. *Opt. Eng.* **1988**, *27*, 276503. [CrossRef]
91. Larson, D.R.; Phelan, R.J. Hydrogenated amorphous germanium detectors deposited onto channel waveguides. *Opt. Lett.* **1990**, *15*, 544–546. [CrossRef] [PubMed]
92. Yi-Yan, A.; Chan, W.K.; Gmitter, T.J.; Florez, L.T.; Jackel, J.L.; Yablonovitch, E.; Bhat, R.; Harbison, J.P. Grafted GaAs De-tectors on Lithium Niobate and Glass Optical Waveguides. In *Proceedings of the 5th European Conference on Integrated Optics: ECIO'89*; International Society for Optics and Photonics: Bellingham, WA, USA, 1989; Volume 1141, pp. 154–159.
93. Le Coarer, E.; Blaize, S.; Benech, P.; Stefanon, I.; Morand, A.; Lérondel, G.; Leblond, G.; Kern, P.; Fédéli, J.M.; Royer, P. Wavelength-scale stationary-wave integrated Fourier-transform spectrometry. *Nat. Photonics* **2007**, *1*, 473–478. [CrossRef]
94. Thomas, F.; De Mengin, M.; Duchemin, C.; Le Coarer, E.; Bonneville, C.; Gonthiez, T.; Morand, A.; Benech, P.; Dherbecourt, J.B.; Hardy, E.; et al. High-performance high-speed spectrum analysis of laser sources with SWIFTS technology. In *Proceedings of the Photonic Instrumentation Engineering*; International Society for Optics and Photonics: Bellingham, WA, USA, 2014; Volume 8992, p. 89920I.
95. Arnaud, L.; Bruyant, A.; Renault, M.; Hadjar, Y.; Salas-Montiel, R.; Apuzzo, A.; Lérondel, G.; Morand, A.; Benech, P.; Le Coarer, E.; et al. Waveguide-coupled nanowire as an optical antenna. *J. Opt. Soc. Am. A* **2013**, *30*, 2347–2355. [CrossRef] [PubMed]
96. High-Resolution High-Rate Laser Spectrum Analyzer. Available online: https://resolutionspectra.com/products/zoom-spectra/ (accessed on 25 April 2021).
97. Hellesø, O.G.; Benech, P.; Rimet, R. Interferometric displacement sensor made by integrated optics on glass. *Sens. Actuators A Phys.* **1995**, *47*, 478–481. [CrossRef]
98. Lang, T.; Genon-Catalot, D.; Dandrea, P.; Duport, I.S.; Benech, P. Integrated optical displacement sensor with four quadrature phase-shifted output signals. *J. Opt.* **1998**, *29*, 135–140. [CrossRef]

99. Bastard, L.; Broquin, J.-E.; Gardillou, F.; Cassagnettes, C.; Schlotterbeck, J.P.; Rondeau, P. Development of a ion-exchanged glass integrated optics DFB laser for a LIDAR application. In Proceedings of the Integrated Optics: Devices, Materials, and Technologies XIII, San Jose, CA, USA, 26–28 January 2009; International Society for Optics and Photonics: Bellingham, WA, USA, 2009; Volume 7218, p. 721817.
100. Verbeek, M.J.; Jentink, H.W. *Optical Air Data System Flight Testing*; NLR: Amsterdam, The Netherlands, 2012.
101. Haguenauer, P.; Berger, J.-P.; Rousselet-Perraut, K.; Kern, P.; Malbet, F.; Schanen-Duport, I.; Benech, P. Integrated Optics for Astronomical Interferometry. III. Optical Validation of a Planar Optics Two-Telescope Beam Combiner. *Appl. Opt.* **2000**, *39*, 2130–2139. [CrossRef] [PubMed]
102. Berger, J.; Haguenauer, P.; Kern, P.; Perraut, K.; Malbet, F.; Schanen, I.; Severi, M.; Millan-Gabet, R.; Traub, W. Integrated Optics for Astronomical Interferometry-IV. First Measurements of Stars. *Astron. Astrophys.* **2001**, *376*, L31–L34. [CrossRef]
103. Olivier, S.; Delage, L.; Reynaud, F.; Collomb, V.; Trouillon, M.; Grelin, J.; Schanen, I.; Minier, V.; Broquin, J.E.; Ruilier, C.; et al. MAFL experiment: Development of photonic devices for a space-based multiaperture fiber-linked interferometer. *Appl. Opt.* **2007**, *46*, 834–844. [CrossRef]
104. Allenet, T.; Geoffray, F.; Bucci, D.; Canto, F.; Moisy, P.; Broquin, J.E. Microsensing of plutonium with a glass optofluidic device. *Opt. Eng.* **2019**, *58*, 060502. [CrossRef]

applied sciences

MDPI

Review

Towards a Glass New World: The Role of Ion-Exchange in Modern Technology

Simone Berneschi, Giancarlo C. Righini and Stefano Pelli *

"Nello Carrara" Institute of Applied Physics, IFAC-CNR, via Madonna del Piano 10, I-50019 Firenze, Italy;
s.berneschi@ifac.cnr.it (S.B.); g.c.righini@ifac.cnr.it (G.C.R.)
* Correspondence: s.pelli@ifac.cnr.it; Tel.: +39-055-522-6393

Abstract: Glasses, in their different forms and compositions, have special properties that are not found in other materials. The combination of transparency and hardness at room temperature, combined with a suitable mechanical strength and excellent chemical durability, makes this material indispensable for many applications in different technological fields (as, for instance, the optical fibres which constitute the physical carrier for high-speed communication networks as well as the transducer for a wide range of high-performance sensors). For its part, ion-exchange from molten salts is a well-established, low-cost technology capable of modifying the chemical-physical properties of glass. The synergy between ion-exchange and glass has always been a happy marriage, from its ancient historical background for the realisation of wonderful artefacts, to the discovery of novel and fascinating solutions for modern technology (e.g., integrated optics). Getting inspiration from some hot topics related to the application context of this technique, the goal of this critical review is to show how ion-exchange in glass, far from being an obsolete process, can still have an important impact in everyday life, both at a merely commercial level as well as at that of frontier research.

Keywords: ion-exchange; optical glasses; glass strengthening; noble metal nanoparticles; rare earths; luminescence enhancement; SERS; flexible photonics; whispering gallery mode microresonators; photovoltaic cell

check for **updates**

Citation: Berneschi, S.; Righini, G.C.; Pelli, S. Towards a Glass New World: The Role of Ion-Exchange in Modern Technology. *Appl. Sci.* **2021**, *11*, 4610. https://doi.org/10.3390/app11104610

Academic Editors: Jesús Liñares Beiras and Giancarlo C. Righini

Received: 25 April 2021
Accepted: 14 May 2021
Published: 18 May 2021

Publisher's Note: MDPI stays neutral with regard to jurisdictional claims in published maps and institutional affiliations.

1. Introduction

To celebrate, over time, the wedding anniversary between ion-exchange and glass, all noble metals (e.g., silver, gold, platinum, etc.) and gemstones (e.g., sapphire, emerald, diamond, etc.) offered by our generous planet would not be enough! In fact, the origin of this lucky marriage is lost in the dawn of time. The art of ancient glassmakers can be traced back to the 3rd millennium BC among the peoples of Egypt and the Middle East where the main raw material, the silica rich sand, was widely available. From there, this knowledge was transferred to the peoples of the Mediterranean Sea, including the Greeks and the Romans (see, for instance, the natron glass used during the Roman Empire period). Over these three millennia, the colouring of glass was given directly during its manufacture by introducing additive metal oxides, such as iron, copper, manganese, etc., to the raw materials before firing the whole system in a kiln and then cooling it in order to obtain the final product [1,2].

Only between the 6th and 7th centuries AD, the Egyptians began to use silver or copper pigments—in the form of powders—to decorate and colour the surface of their artefacts, such as dishes, vessels and pots, by means of a thermal annealing process in a furnace. Through this technique, the pigment became part of the atomic structure of the products, colouring them in correspondence of their surface. However, due to the low temperature of the firing process and the difficulty of controlling it, the final artefacts remained rather opaque [3,4]. Later, around the 9th century AD, this staining method—improved in terms of technological process (i.e., higher operating temperature; better kiln performance) and now capable of giving glass lustre to ceramic artefacts—found great diffusion among the Mesopotamian peoples before arriving in Spain, in the first centuries of the millennium, following the expansion of Islamic culture [5,6]. In Europe, ion-exchange experienced a notable application during the outbreak of Gothic architecture (between 12th and 14th century) contributing to the realisation of the wonderful multi-coloured stained-glass windows of the cathedrals of that period. A paste, composed of clay/ochre and silver chloride/sulphide, was spread on the glass to be treated in order to produce a thin layer on its surface. Subsequently, the whole system was fired in a reducing atmosphere—induced in the furnace by introducing smoking substances—at a temperature close to the softening point of the glass. In these conditions, the ion-exchange process was triggered: the silver ions diffused inside the glass replacing the alkaline ions, such as sodium, widely present in the pristine glass. Finally, the simultaneous presence of reducing elements, both external (i.e., the smoking elements in the atmosphere) and internal to the glass (i.e., impurities, such as iron or arsenic), favoured the formation of silver metal nanoparticles (NPs) inside the ion-exchanged specimen, leading to a yellow-amber colour whose tonality and intensity depending on the process parameters (i.e., time and temperature) and the size of the NPs thus formed (from a few to one hundred nanometres), respectively. Other colours, such as red or green, were obtained using iron and copper salts, respectively. In the following renaissance period, this decorative technique became particularly popular in Italy in the creation of polychrome and lustre pottery artefacts, among which those of Deruta stood out par excellence [5,7].

It was only around the beginning of the 20th century that the foundations for a possible application of the ion-exchange process in the technical-industrial field were laid. In 1913 G. Schultze was the first to study the diffusion of silver ions into the glass using silver nitrate salt ($AgNO_3$) as ion source, starting a whole series of studies aimed to understand the chemical-physical nature of the phenomenon and its effects on some physical properties of the glass so treated. In particular, a few years later, in 1918 at the Schott Glass Laboratory, it was demonstrated that ion-exchange produces an increase of the refractive index of the layer of the glass involved in the diffusive process [4,8]. Only from the 1960s, however, the ion-exchange technique became an industrial standard process.

Among all the possible applications, glass strengthening represented the first example of glass modification by this standardized technique. The process was initially investigated in 1962 by S. S. Kistler and P. Acloque (the latter in collaboration with J. Tochon) who, in their works, demonstrated how the in-diffusion of ions having different atomic size in the pristine glass matrix (i.e., larger potassium ions K^+, originally present in the molten salt KNO_3, in place of smaller sodium ions Na^+ in the glass) was able to prevent or heal over the possible formation of micro/nano-cracks on the specimen surface, increasing its mechanical strength [9,10]. Since then, many efforts have been carried out until our days in this field both at research and industrial levels, and the development is still under way with amazing perspectives [10–13].

Almost simultaneously to glass strengthening, another key application of ion-exchange technique quickly began to establish itself. In fact, at the turn of '60s and '70s, the availability of the first optical fibres by Corning Inc (Corning, NY, USA), joined with the revolutionary vision of Bell Laboratories researchers J.E. Miller and colleagues, led to the aim of miniaturizing and integrating all necessary components for the generation, addressing, processing and detection of an optical signal on a single chip by means of optical waveguides and gave the decisive impulse for the development of integrated optics (IO). The ion-exchange process in glass greatly contributed to achieving these results [14–17]. Pioneering works in this field were given by T. Izawa and co-workers and T. G. Giallorenzi et al., at the beginning of the 1970s, who demonstrated the feasibility to obtain planar glass waveguides by thermal or field assisted ion-exchange using different molten salts, thus opening up the way for further development of novel integrated devices in the optical communications field [18,19]. Optical waveguides in glass by ion-exchange technique offered several advantages in comparison with other fabrication methods and materials: compatibility with optical fibres, low propagation and coupling losses, low birefringence, low cost in terms of material and fabrication process, high stability and reliability. All these features contributed in the following decades to the boom of IO devices with different functionalities, such as passive (e.g., for signal addressing and processing), active (e.g., for signal generation or amplification, thanks to rare earths doping), or even hybrid, when both functionalities coexist in a same chip [20–39].

The third noteworthy application of ion-exchange was historically linked to the glass staining technique by the introduction of metal ions inside the pristine amorphous matrix during the process or, more frequently, by inducing their subsequent reduction into metallic form through an ad-hoc thermal post-process [40,41]. However, this approach of loading glass with metal ions or nanoparticles has been shown to go far beyond the mere colouring of the material. In fact, around the last decade of the 20th century, the progress in plasmonics—with the possibility to obtain both strong local field enhancement and high absorption cross-section by means of the surface plasmon resonance phenomenon in ion-exchanged dielectric substrates with embedded metal nanoparticles—has rapidly increased the research interest for these nanostructured materials [42–47], broadening their application horizon.

For instance, it was demonstrated that in ion-exchanged glasses the presence of noble metal nanoparticles (e.g., Au, Ag, or Cu), with their large third-order nonlinear susceptibility and ultrafast response, makes these materials suitable for applications also in non-linear optics [48–53].

Moreover, these embedded noble metal ions and nanoparticles played—and still do—an important role in the luminescence properties of the same ion-exchanged host materials especially when these are doped with elements belonging to the rare earth group, mitigating the limitation imposed on their concentration due to the occurrence of quenching phenomena. From the early works of Malta et al. [54] and Hayakawa et al. [55], the research literature on this topic has highlighted different reasons concerning the origin of the rare earth luminescence enhancement in optical glasses chemically treated by ion-exchange, depending on the nanostructure sizes so realized: (i) an energy transfer (ET) process between the energy levels of the noble metal ions (such as isolated Ag^+, Ag^+ - Ag^+ pairs or silver aggregates) and those belonging to rare earths ions; (ii) an ET mechanism among non-plasmonic small metal nanoparticles (i.e., molecule-like nanoparticles having few nanometres size) and the doping active elements; (iii) a local field enhancement around the rare earth ions due to the surface plasmon resonance (SPR) phenomenon induced in small metal nanoparticles [54–66]. In the technological field, the aforementioned mechanisms—together with the down/up-conversion ones—have favoured the use of these ion-exchange glass systems in the photovoltaic sector in order to increase the solar cell efficiency through tailored cover-glasses and in the development of low cost and high performance white/coloured solid-state light sources [67–81].

On the other hand, in sensing/biosensing applications, low-loss ion-exchanged planar waveguides with embedded metal ions represented the core of integrated optical sensors for the analytical determination of biomolecules at the interfaces by an evanescent wave interrogation mechanism, exploiting different optical detection approaches such as those based on fluorescence or absorption [82–90]. Furthermore, these ion-exchange optical waveguides have been used as excitation systems for surface plasmon resonance (SPR) or localized surface plasmon resonance (LSPR) in noble metal thin films or NP arrays selectively deposited on them. This approach replaced the more traditional Kretschmann configuration based on prism coupling and contributed to the realisation of hybrid-plasmonic label-free integrated optical biosensors with high performance [91–94]. Alternatively, the same noble NPs can be directly grown in the guiding structure in order to obtain LSPR optical waveguides for the detection of selected analytes by monitoring the changes in absorption spectra as a result of the chemical reaction occurred at the interface [95]. Last but not least, the possibility of making optical waveguides in one-dimensional (1D) or two-dimensional (2D) geometry with embedded noble metal nanoparticles, starting from glass substrate by means of ion-exchange technique and an ad-hoc thermal post-process, has contributed in the last decades to the strong development of planar platforms for sensing/biosensing applications based on surface enhanced raman scattering (SERS) mechanism [96–104]. In comparison to the chemical procedures commonly adopted to bind metal NPs to the surface of a sample, the ion-exchange technique followed by a suitable thermal post-process represents a simpler and less expensive method for the manufacture of SERS substrates and provides the not negligible advantage of long-term stability of the metal nanoparticles thanks to their embedding within the same glass specimen [105–107].

From this wide range of scenarios, getting inspiration from some hot topics related to the application context of this technique, the goal of this critical review is to show how ion-exchange on glass, far from being an obsolete process, can still have an important impact in the everyday life, both at a merely commercial level as well as at that of frontier research.

2. Glass Strengthening: The Ion-Exchange Contribution

Today, it has become a cliché to associate the idea of glass with that of brittleness, so that, in the course of daily speaking, it is common to use the expression "fragile as glass" whenever one refers to an object destined to easily break. In this regard, Tennessee Williams, one of the most popular American playwrights, expresses himself in these terms about glass: "When you look at a piece of delicately spun glass you think of two things: how beautiful it is and how easily it can be broken" [108]. Moreover, the idea of brittleness associated with glass is so rooted in our mind that it stands as symbol of the uncertainty of life and the human condition as expressed by G. K. Chesterton in one of his aphorisms: "I felt and feel that life itself is as bright as the diamond, but as brittle as the window-pane" [109].

Yet, in spite of these beliefs, a pristine glass can tolerate high stresses, of the order of GPa. This makes glass an extremely strong material in itself [11–13,110]. A classic example is given by an optical fibre which, despite its size as thin as a hair, can achieve extraordinary mechanical tensile strengths (~few GPa) [111].

So, what does determine glass brittleness? The answer lies in the presence on its surface of microscale flaws occurring during the different steps of its manufacturing, processing and handling. Consequently, glass strength is drastically reduced by several orders of magnitude, down to only a few tens of MPa before reaching breaking point, once a mechanical stress is applied to its surface. If fused silica glass, thanks to its intrinsic structure, presents the best characteristics in terms of hardness and mechanical strength, multicomponent oxide glasses, such as soda-lime or float ones, are among the most produced due to their low cost and easier fabrication. However, due to their highly heterogeneous composition, they are more liable to the formation of surface defects during their processing and, therefore, more prone to breakage once they are subjected to mechanical stress. Unlike crystals which have a well-ordered internal atomic structure, glasses are characterized by a disordered and chaotic molecular configuration typical of the amorphous materials. The absence of a crystal lattice leads to the formation of shear bands in the material. From the observation of these bands that occurred in a metallic glass following their manufacturing process, A. Wisitsorasak and P. Wolynes developed a general theory capable of explaining how the breaking phenomenon occurs in these materials [112]. The model provides a mapping of all the possible configurations of the molecules in the solid, in order to describe how the mechanical stress changes the atomic rearrangement rate in the glassy material, with the consequent formation of the shear bands. The observation of these bands allows to identify the crystallisation points where the glass is structurally weaker and, therefore, could be more subject to breakage when exposed to external stress. The model can be extended and generalised for any other glass formulation. In this way it is possible to produce glass with better mechanical characteristics, which respond to different needs depending on their use. In addition to this promising theoretical model, currently, the main practical techniques for strengthening glass involve the use of thermal treatments or chemical tempering of the material [11–13].

2.1. Thermal Strengthening

Annealed, heat strengthened and chemically strengthened (or tempered) glass are the three main types of glass for general use. Annealed glass is the "regular" glass, as it comes from industrial glass manufacturing (float glass). Thermal strengthening involves a rapid cooling in air of the annealed glass surface after it has been heated to a temperature above the glass transition temperature T_g and close to the melting point of the material. At the beginning of this cooling process the glass surface cools down and contracts quickly, while its internal core remains hot in order to compensate dimensional changes with small relaxation stress. In this condition, the inner region of the glass is subject to compression while the external one is in tension. When the glass interior cools and contracts, the surfaces are already rigid and therefore residual tensile stresses are created inside the glass while compressive stresses occur at the surface. In comparison to a simple annealed glass, this thermal tempering process increases the strength of the pristine material because the applied stresses must overcome the residual compressive ones on the surface, thus preventing the breaking event [13,113,114]. This allows to increase the mechanical strength of the glass by about 4–5 times (up to a few hundred of MPa) and, at the same time, guarantees a better behaviour of the material in terms of safety when it is subject to a breakage [115]. In fact, when this event occurs, this kind of glass generally shatters into many small non-sharp and quite harmless fragments unlike what happens for annealed glasses which break in larger pieces, with a dangerous dagger-like profile. This particular response of tempered glass to a breakage event can be explained by the high strain energy present within these materials. The main limitation of the thermal strengthening process is linked to the minimum thickness obtainable for a tempered glass, which cannot reach values lower than a few millimetres (around 2–3 mm, typically) due to the rapid cooling process and the substantial difference in cooling rate from the glass core to its surface. This inevitably imposes further limits on the geometry and shape of the final products made with this type of glass, which must be simple and not very elaborate [11–13,67]. For all these reasons, thermally tempered glass finds its widest use in all those applications where human safety becomes an issue, such as in the automotive sector, for the development of side and rear windows in the vehicles, in the architectural one for the realisation of safety glass doors in private and public buildings, in the domestic environment (see, for instance, the glass for showers or microwave ovens), and the military/civil sector for the development of bulletproof glass where tempered glass is an essential component.

2.2. Chemical Strengthening

Another method to increase the strength of a multicomponent oxide glass involves modifying its chemical composition, replacing one of its constituent elements with another one that is different in terms of atomic dimensions and electronic polarizability and, at the same time, without significantly changing the network structure of the pristine glass. In particular, the replacement of small ions (originally present in the glass matrix) by larger ones (coming from an external ion source) produces a compressive stress at the surface of the glass for a "stuffing" or "crowding" effect, which can improve the glass strength by counteracting the surface flaws. This compressive stress at the surface is compensated by a tensile one in the internal region of the glass [7,11–13,110]. Generally, this method is applied at a temperature lower than the glass transition temperature T_g and involves the use of a mixture of molten salts as an ion source. For this reason, the "chemical strengthening" method is also referred to as "ion-exchange strengthening". This technique prevents any risk of surface deformation and geometrical distortion due to the temperature parameter if compared to the thermal strengthening method. The resulting compressive stress so generated is generally higher than that achievable with the thermal tempering approach and it is confined in a thinner region of the glass, immediately below its surface, for low process temperature. In this case, the maximum compressive stress obtainable with this technique ranges from several hundreds of MPa to around 1 GPa in proximity of the glass surface [110,116]. Conversely, for a high process temperature (i.e., close to the glass T_g or above), the position of this maximum is located several microns below the glass surface and the magnitude of the stress rate decreases for the occurrence of surface stress relaxation due to viscous flow of the glass [13]. On the other hand, for a same process temperature, the maximum of the surface compressive stress decreases when the ion-exchange time increases: the depth at which the residual stress equals zero (i.e., the depth of layer, DoL) becomes deeper and, consequently, the flaws that can be involved in the process are larger [110,117,118]. On the other hand, the maximum for the tensile stress is not so easy to identify but, in some cases, it is located immediately below the compressive zone and its value is an order of magnitude lower than that of the compressive stress (~few tens of MPa), depending on the process duration and sample thickness [13]. Figure 1 compares the residual stress profile in the case of thermally and chemically toughened glass, respectively. Due to the stuffing effect of the incoming ions at the glass surface, the chemical strengthening method presents both large surface compression and internal tensile stress. The transition from the compression zone to the tension one is quite sharp. Conversely, the residual stress distribution in a thermal tempered glass takes on a parabolic profile depending on the temperature distribution achieved during the cooling step of the thermal tempering process. In this case, the compressive stress at the surface is approximatively twice the value of the tensile stress inside the specimen [11–13].

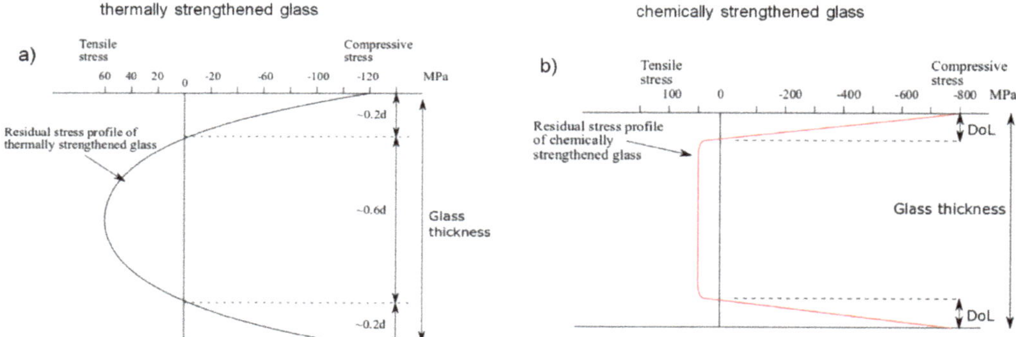

Figure 1. Residual stress profile through the glass thickness for: (**a**) thermally tempered glass; (**b**) chemically toughened glass. d is the glass thickness; DoL is the depth of layer. Figure reprinted from [67] under the terms of the Creative Commons Attribution 4.0 International License.

Several approaches have been proposed for stress evaluation in chemically toughened glasses. Generally, they are based on Cooper's method which provides the determination of the stress in the glass starting from the knowledge of the concentration gradient for the ions participating in the exchange process [7,119]. In this context, R. Dugnani developed an analytical solution to the problem assuming a generalised function for the stress relaxation, a constant ionic inter-diffusion coefficient and, simultaneously, taking into account an analytical approximation for the composition-dependent stress relaxation behaviour of the glass [120]. On the other hand, A. K. Varshneya, G. Macrelli, and others in their studies modified the Cooper's analysis introducing a new term in the model, related to different relaxation contributions (i.e., the viscoelastic and structural one) together with the network hydrostatic yield strength in order to evaluate the subsurface maximum compressive stress when the process temperature is higher and/or close to the glass transition temperature T_g [121–124]. Also noteworthy is the method, recently proposed by R. Rogoziński, which allows to control in real time the stresses generated in a glass as a result of the ion-exchange process from the knowledge of some parameters such as the elasto-optical coefficients, the dependence of the diffusion coefficients on the temperature and, finally, the function describing the time relaxation of stresses at the glass surface [118]. Differently from the numerical and analytical approaches mentioned above, N. Terakado and co-workers have developed a very original method for evaluating the compressive stress in a chemically strengthened glass, connecting it directly to the glass structure on an atomic scale. The non-contact and non-destructive method is based on the "stuffing" effect and the knowledge of three structural parameters, represented by specific boson peaks in the micro-Raman spectra of the sample [125].

At the end of this brief overview on the main concepts underlying the chemical strengthening of a glass, it should be remembered that in addition to the two most important process parameters, time and temperature, two other factors influence the formation of stress in a chemically strengthened glass: the composition of the pristine glass and that of the molten salt solution in which it is immersed. In fact, the larger the difference between the atomic dimensions of the two ionic species involved in the exchange, the greater is the magnitude of the compressive stress near the surface. In this sense, sodium-potassium ($Na^+ \leftrightarrow K^+$), lithium-sodium ($Li^+ \leftrightarrow Na^+$), or lithium-potassium ($Li^+ \leftrightarrow K^+$) exchanges are able to produce high compressive stresses with values that, practically, are close to 1 GPa in the best cases [11–13]. On the other hand, the composition of the pristine glass plays an important role on the diffusion rate of the ions and stress relaxation of the material during the exchange process. The common soda-lime glass has excellent diffusivity values for the incoming ions but a rapid relaxation process which makes difficult the achievement of a good glass strengthening despite its high alkali content [126]. However, the most promising glass formulation is the one that refers to aluminosilicates which have the highest diffusivity values while ensuring the achievement of compressive stresses well above the possible relaxation effects of the glass matrix [127,128]. In particular, the closer the alumina content is to that of the alkaline oxides, the higher the attainable strength of the glass following the exchange process [129]. For more details on the different characteristics of glass formulations, molten salt compositions, and ion-exchange typologies in glass strengthening, the reader is referred to the exhaustive reviews on the subject [11–13]. The main features of the two glass strengthening processes described above are compared in Table 1 below.

Table 1. A comparison between thermal and chemical glass strengthening methods.

Parameter	Thermal Strengthening	Chemical Strengthening
Strength (max. value range)	~(200–400) MPa	~(800–1000) MPa
Surface compression layer	Thick	Thin
Stress distribution profile	Parabolic	More flat & square
Minimum sample thickness	2–3 mm	<1 mm
Process Time	Short (mins)	Long (hours)
Process Cost	Cheap	Expensive
Glass Composition	No relevant	Alkaline glass (i.e., soda-lime or aluminosilicate)
Product Shape	Simple geometries	Unusual shape

2.3. Chemical Strengthening: Applications

Although the chemical toughening process of glass presents expensive production costs due to its high time consuming if compared with a simple thermal tempering process, the possibility of obtaining greater mechanical strengthening in large glass surfaces with small thicknesses makes this process particularly suitable for many technological applications, ranging from the architectural (e.g., windows for buildings and palaces), automotive and transport sectors (e.g., windshield for airplanes and vehicles) up to the electronics and military ones (e.g., panels for displays in many electronic devices; air-to-ground missile launch tube protective covers). Figure 2 shows some outstanding applications related to chemical strengthened glasses.

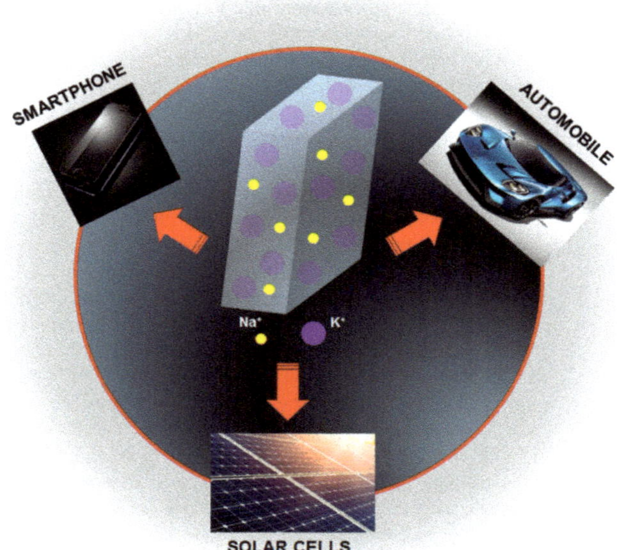

Figure 2. Some relevant application fields of glass chemical strengthening.

In daily life, storing and sharing of data, from those personal to public ones, is almost delegated to electronic devices, such as smartphones, tablets, and lap-tops. The safety of these data does not only concern their protection at software level from any external attacks or a sudden failure of the operating system, but also requires the physical protection of the same devices, with particular reference to some of their most vulnerable elements such as monitor and/or display.

For this reason, CORNING® Inc., Asahi Glass Corporation (AGC), and SCHOTT AG, three world leaders in glass manufacturing, have developed in the recent years different high ion-exchange (HIE) tempered aluminosilicate glasses (for instance, the Gorilla® series from CORNING®, the Dragontrail™ from AGC, and the Xensation series from SCHOTT AG, respectively) with outstanding resistance to breakage and scratches in order to increase the protection for cover and touch screens, greatly reducing the risk of damage due to potential impact or fall and, consequently, avoiding any repair costs associated with this unpleasant event [130–133].

In all these cases, the characteristics of high surface resistance (>900 MPa in the best cases), flexibility (>700 MPa as a bending stress) and lightness (glass thickness < 1 mm, generally) are conferred to the pristine aluminosilicate glasses by the special chemical tempering treatment based on $Na^+ \leftrightarrow K^+$ ion-exchange, where the large K^+ ions, substituting the host Na^+ ions, force the generation of surface compression stress, supporting the closure of the cracks, thereby increasing the strength of the material and preventing the formation of others.

In this context, Apple, although employing Corning Gorilla glass technology to protect the displays of its smart devices, in recent years made a strong effort in the search of alternative solutions to further increase the conventional method of glass strengthening. Different strategies have been investigated, such as the implementation of a patterned asymmetric method based on double ion-exchange process in order to increase the compression depth over at least one localized region, with particular reference to the bend zones of the glass cover, as highlighted in some Apple patents recently published by the US Patent &

Trademark Office [134–136]. Regarding the bulk material, the choice of the manufacturing company fell on a lithium aluminosilicate glass for its excellent chemical durability and mechanical resistance. The whole toughening process consists of two ion-exchange steps. In the first one, the specimen is immersed in a sodium nitrate $NaNO_3$ salt (concentration range: 30–100% mol.) at a temperature lower than the T_g of the glass (process temperature range: 350–450 °C) for 4–5 h. In this case, the sodium ions, originally present in the melted salt, may exchange for smaller lithium ions in the glass matrix. This contributes to a first glass strengthening, allowing a build-up of sodium ions at the specimen surface. The second ion-exchange step, carried out in a potassium nitrate KNO_3 salt (concentration range: 30–100% mol.) at a temperature lower than the glass T_g (process temperature range: 300–500 °C) for a process duration of 6–20 h, contributes to the definitive toughening of the glass. This double ion-exchange process was adopted by Apple both for the realisation of the front and rear glass cover of the iPhone11 smartphone generation. Moreover, the rear glass cover is also loaded with metal elements—through diffusion or deposition processes—in order to guarantee wireless charging operation for the final device.

Corning response to Apple arrived quickly with its Gorilla® Glass Victus™, a chemically treated aluminosilicate glass capable of guarantee the best performance in terms of drop (up to 2 m) and scratch resistance among all the glasses belonging to the Gorilla® family. The Gorilla® Glass Victus™ presents a minimum thickness of about 0.4 mm and a shear modulus equal to 31.4 GPa, fully responding to the need for high mechanical strength in small thicknesses [137]. Due to these remarkable features, this cover glass was adopted by Samsung in some of its products such as, for instance, the Galaxy Note 20 Ultra.

Finally, it should be mentioned that the race for increasingly resistant materials is moving towards the use of glass-ceramics, characterized by the presence of nanocrystals within the amorphous glass matrix. In fact, a glass-ceramic is born like glass and then turns into an almost completely crystalline substrate by an ad-hoc thermal treatment able to induce the nanoparticle formation from additional nucleating agents, such as silver or titanium, previously introduced in the glass matrix. The dimensions of these nanostructures are generally of the order of a few nanometres, thus allowing an excellent transparency of the material. Moreover, glass-ceramics are tough, lightweight and present high temperature stability, high resistivity and excellent isolation capabilities. All these features are well suited to the role that these materials must assume in terms of safety of smart electronic systems. It is therefore no coincidence that, at the release of the new iPhone12 series in October 2020, Apple declared that it had realized—in collaboration with Corning—a new extremely resistant glass-ceramic product, the Ceramic Shield, four times better in terms of drop performance than Gorilla® Glass Victus™ [138]. Although Apple is not the first to have used glass-ceramic materials for the protection of its smart electronic devices (see, for instance, the Samsung with its smartphone Samsung Galaxy S10), the company from Cupertino is however the first to use such materials for the display protection while other manufacturing companies, such as Samsung, have used this material for back panel or camera-lens safety.

The chemical strengthening process of glass has recently found its application also in the automotive sector. In 2017, the American automotive manufacturer Ford was the first to equip its flagship automobile, the Ford GT supercar, using Gorilla glass for the windshield, rear window and engine cover of the vehicle, replacing the use of more traditional tempered glass. The new Gorilla hybrid glass, designed in strict collaboration with Corning, allowed a lightening of the overall weight of the automobile by well over 5 kg, increasing its manoeuvrability and reducing fuel consumption as well as the risk of glass breaking. Instead of using two thermally tempered sheets, joined together by a transparent thermoplastic adhesive in order to form the classic windshield glass, the new solution uses the co-presence of three layers: an external tempered glass joined to an internal one, highly resistant and chemically toughened, by a transparent and sound-absorbing thermoplastic. The presence of the ion-exchanged glass contributed to reduce the overall thickness of the new hybrid glass by approximately 50% compared to a common windscreen made with thermally tempered glass. Given the advantages of this strategy, it is likely that in the future the same technology will be extended to the production of series cars [139,140].

As a third significant example of ion-exchange application for glass strengthening, the case of photovoltaic and space solar cells deserves consideration. In fact, the cover glass represents one of the fundamental components of each of these devices and plays a key role in the final cost and efficiency of the cells. In terms of cost, the contribution relating to the protective glass is not negligible because it alone represents about 25% of the total cost of the whole device. Moreover, regarding the cell efficiency, the protective glass can influence the device performance depending on the transparency level of its material to solar radiation. Therefore, the optimisation of the cell cover glass becomes extremely important in order to minimize costs and maintain high performance for the final product.

In principle, glasses for photovoltaic and space solar cells should be as lightweight as possible in order to reduce material thickness and waste, so decreasing the overall cost of the device. Usually, commercial, thermally toughened float or soda-lime glasses, with a thickness of around 3 mm, are adopted on-board of the photovoltaic cells while they appear quite bulky and heavy for space solar cells. As a direct consequence of using a thin cover glass, comes the need for greater strength of the material. Moreover, the glass composition should be engineered for low losses and small UV photon absorption in order to guarantee better cell efficiency and service lifetime, respectively [67]. The use of chemically strengthened glasses may represent a valid solution, as reported by H. Wang et al. [141]. In that work, the authors demonstrated how an ion-exchange process, performed in a bath of pure KNO_3 molten salt at a temperature comprised between 400 °C and 460 °C for a process time from 20 to 90 min, was able to confer suitable toughening to experimental silicate glasses having a thickness of only 120 μm. In particular, the maximum achievable flexural strength was measured to be around 632 MPa, four times higher than that of the un-treated samples. Moreover, the selected ionic process did not significantly modify the transparency characteristic of the pristine glasses in a wide range of the wavelength, while, at the same time, it was able to increase the durability of the material when exposed to the ion radiation. Hence, the obtained results were in full agreement with the possibility to reduce the cost and improve the performance of the space solar cell in terms of efficiency and service lifetime.

With the aim to release on the market a lighter glass, the researchers of the Asahi Glass Corporation (AGC) made a novel chemically-strengthened industrial glass, Leoflex™, an ultra-thin, flexible and extremely resistant material. Leoflex™ glass, with its lower sodium concentration at the surface, exhibits better electrical resistivity and higher mechanical strength caused by the chemical toughening process and strongly reduces the risk of potential induced degradation (PID), which represents one of the main reasons for the failure of photovoltaic and space solar cells. All these features made this material an ideal candidate as a cover glass for the development of highly performing photovoltaic application [142]. Additionally, the ion-exchange process also plays a key role in the efficiency of the same cell by introducing noble metal ions/nanoparticles capable of triggering particular energy transfer mechanisms (i.e., down-conversion from ultraviolet (UV) to visible wavelengths or up-conversion from infrared (IR) to visible range), especially when the glasses are doped with elements belonging to the rare earth group. This subject, together with other ones which involve the use of metal aggregates for different application fields, will be addressed in the next section.

3. Noble Metal Ions/Nanoparticles by Ion-Exchange in Glass Substrates: Novel Platform for Photonic & Sensing Applications

3.1. Ion-Exchanged Noble Metal Ions/Nanoparticles in Rare-Earth—Doped Glasses for Enhanced Light Sources and Photovoltaic Cells

The development of solid-state light sources by using glasses doped with different combinations of rare earths has driven the attempts to further improve their efficiency by using sensitisers added to the glass matrix. Their function is to increase the absorption of the excitation light, also at wavelengths where more efficient, reliable, and less expensive pump sources are available, and efficiently transfer the collected energy to the emitting rare earths.

The same materials can also be tested to be used to enhance the efficiency of photo-voltaic (PV) cells.

In fact, a large portion of solar radiation in the UV spectrum cannot be absorbed or efficiently converted in electricity by silicon-based PV systems. In case the solar light in the spectral windows where PV cells are less efficient could be converted entirely inside the working wavelength interval of the cells, their theoretical efficiency would increase from 29% to 38%. It is therefore apparent that even a more realistic partial result could be a crucial improvement.

Many approaches have been pursued, but, as we will describe in the following, usually metallic silver nanoparticles have been used as sensitizer. Their advantage is a strong broadband absorption band, joined to the possibility of introducing them in the glass by means of ion-exchange and subsequent thermal processes, which makes this mature technological approach fairly convenient and economical.

Malta et al. reported, in 1985, the enhancement of Eu^{3+} luminescence in melted calcium boron oxy-fluoride glass, which they related to the local field enhancement produced by the surface plasmon resonance of the Ag particles [54]. In this case, silver was introduced directly in the glass melting process. This result was confirmed by Hayakawa et al. in sol-gel glasses where reduction of the silver ions to metallic form was obtained by annealing the glass in H_2/N_2 atmosphere [55]. Similarly, Li et al. demonstrated for the first time non resonant energy transfer from Ag^+ ions (hence, non-metallic) to Dy^{3+}, Sm^{3+} and Tb^{3+} [143].

These are only few of many examples of rare earth doped glasses where ion-exchange was not used as a means for introducing silver nanoparticles in the glass and that showed the potential of using silver as sensitiser. In the following, we will focus on and limit to examples where ion-exchange was applied.

Jiao et al. investigated the effects of silver introduced by ion-exchange in sodium borate glasses on Eu^{2+} and Eu^{3+} ions fluorescence [144]. The authors stressed the different role that few Ag atom/ion complexes, up to full grown nanoparticles, may play in influencing fluorescence emission. Actually, after introducing silver in the glass, typical emission of both Eu^{2+} and Eu^{3+} ions increased, but excitation would switch from the 270 nm band of Ag^+ ions to the 350 nm ascribed to Ag clusters, which get formed at expense of Ag^+ isolated ions and smaller aggregations by increasing ion-exchange time. Using 350 nm excitation, an otherwise weak band assigned to the $4f^6 5d \rightarrow 4f^7$ transition of Eu^{2+} ions was also enhanced. Therefore, it is evident how in this case both the optimal excitation and the emission depend on the valence and aggregation of silver, which can be controlled by the ion-exchange process parameters.

Similarly, Li et al. studied the effects of silver doping by ion-exchange on the photoluminescence emission in Eu^{3+} sodium-aluminosilicate glass [60]. As shown in Figure 3, the inclusion of silver aggregates produced a broad emission band in the 400-700 nm interval overlapping the Eu^{2+} 5d-4f emission, and a substantial increase of the 615 nm line of Eu^{3+} when excited at 350 nm as the silver content was increased.

The effect was ascribed by the authors solely to energy transfer from silver ions to Eu^{3+} ions. The combined emission by silver, when suitable ion-exchange parameters were chosen, and Eu^{3+} ions produced white light. Therefore, silver contributed both by enhancing Eu^{3+} emission in the red and by its own emission in the blue-green region. The same group of authors has obtained very similar results with Sm^{3+} and Tb^{3+} in the same glass [59,145] and in borosilicate glass [64].

Strohhöfer and Polman demonstrated energy transfer from silver ions/dymers to erbium ions implanted in BK7 after Ag^+/K^+ ion-exchange [56].

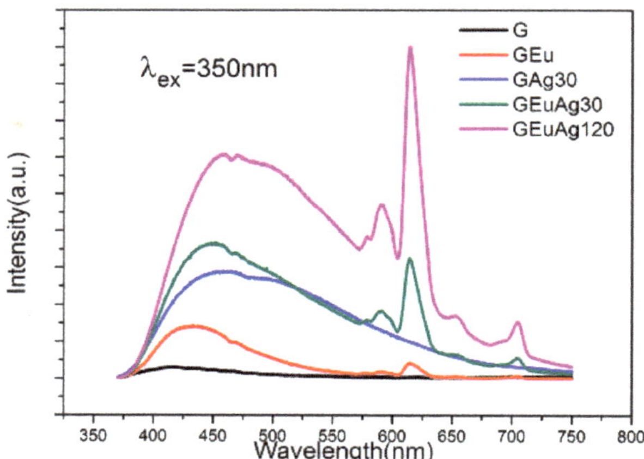

Figure 3. Photoluminescence emission of Eu^{3+} doped sodium-aluminosilicate glass as a function of the silver ions density. Reprinted with permission from [60] © The Optical Society.

Our research group has characterised very thin wafers of Er^{3+}-doped silicate glasses entirely ion-exchanged in an $Ag/NaNO_3$ melt. [57,58]. We tested several post-exchange annealing processes evidencing the effect on absorption and emission spectra. In particular, since Er^{3+} resonant excitation was not affected by the silver presence, whereas non-resonant was enhanced (up to 100 times), we concluded that the effect should be ascribed to energy transfer from the silver nanoparticles to the Er^{3+} ions, rather than local field enhancement.

Similar results have been demonstrated in different Er^{3+}/Yb^{3+} glass formulations, such as silicate and titanoniobophosphate glasses, where planar waveguides have been fabricated by Ag^+/Na^+ ion-exchange [62,146].

Photoluminescence enhancement of Dy^{3+}/Eu^{3+} co-doped tellurite glasses was reported by Yu and co-workers [78]. In this case, the quantum efficiency improvement (several % units, depending on excitation wavelength, ion-exchange conditions and glass doping) was attributed mainly to the local plasmon resonance field enhancement produced by the silver nanoparticles. Also in this case, different ion-exchange parameters produce glasses capable of emitting with different CIE (Commission Internationale de l'Éclairage) coordinates, and therefore emission tailoring is readily possible.

Recently, the Yb^{3+}/Ho^{3+} doping system, interesting for the fabrication of laser sources for medical applications operating at 2 μm, has been studied by Vařák et al. in zinc-silicate glasses [147]. In this case, Ag^+/Na^+ and Cu^{2+}/Na^+ ion-exchanges have been tested, in order to introduce either silver or copper nanoparticles inside the glass. In both cases, planar waveguides were obtained and characterised. Regarding photoluminescence, unfortunately the effect of copper was, even if only slightly, negative, since the emission decreased after ion-exchange, and a subsequent annealing step did not improve it. The effect of silver was however better, and a 20% photoluminescence intensity increase, assigned to energy transfer from silver clusters to Ho^{3+} ions, was measured.

Silver ion-exchange has been reported to enhance NIR emission also when not using rare earths, as with Bi-doped silicogermanate glasses [148]. In fact, lower valence Bi ions are responsible for NIR fluorescence, so that increasing their presence can improve the low efficiency of fibre amplifiers based on this element. In this case, the effect of silver ions introduced in the glass by ion-exchange followed by thermal annealing is not linked to energy transfer processes, but rather to the valence reduction effect on the Bi^{3+} ions. This can be efficiently achieved only by introducing silver ions by ion-exchange, since the required valence state of silver is given by the process itself. The authors showed that Bi emission in the visible range was reduced and NIR emission increased by about twofold this process.

However, the presence of silver does not always enhance photoluminescence emission. In fact, as described for Ce-doped soda-lime silicate glasses by Paje et al. [149], silver causes the reaction:

$$Ce^{3+} + Ag^+ \rightarrow Ce^{4+} + Ag^0.$$

This in turn not only changes the emission photoluminescence spectrum but also its intensity, since Ce^{4+} ions do not emit and even absorb both the UV excitation and the photoluminescence produced by Ce^{3+} ions.

In the case of photo-thermo-refractive (PTR) glasses, however, the same type of reaction has been exploited by Sgybnev et al. [74] to test the potential of these sodium-zinc-aluminosilicate glasses co-doped with Ce and Sb to serve as wavelength converters for enhancing the efficiency of PV cells. The authors tested the possibility of producing neutral silver molecular clusters in PTR glasses both by irradiating with UV radiation samples already doped with silver in bulk or by Ag^+/Na^+ ion-exchange. In both cases an annealing step followed the first phase of the process.

As a result, in all cases, silver molecular crystals were obtained, but their efficiency was different depending on the process. The PV cells placed behind the best sample obtained by ion-exchange reduced their overall efficiency, even if only slightly, whereas the sample with silver atoms dispersed at the time of initial melting showed a slight efficiency improvement respect to the as-synthesised cover glass.

3.2. Ion-Exchanged Noble Metal Nanoparticles in Glasses for Sers Applications

The interaction between light and noble metal nanoparticles (NP) at the material interface is the key point for the employment of one of the best performing optical investigation methods: the surface-enhanced Raman scattering (SERS) technique. These metallic nanostructures work as "hot spots" for the electric field of an interacting lightwave by means of the localised surface plasmon resonance (LSPR) effect [150]. When a molecule is close to one of these "hot spot", its Raman signal increases by several order of magnitude under the action of the enhanced field. This confers to the sensors/biosensors based on this optical technique ultra-sensitive detection limits, combined with the structural information content typical of the Raman spectroscopy [151].

For the development of high-performance SERS devices, the optimisation of the substrates is mandatory. Ideal SERS platforms should ensure high efficiency (in terms of signal enhancement), reproducibility in response, reusability, long-term stability for the metal nanoparticles, low cost, and simplicity of the fabrication method [97,152]. Furthermore, considering that silver NPs—which offer the best SERS activity among all noble metal nanostructures—are prone to oxidation and sulfidation phenomena, it is easy to understand how obtaining SERS substrates adhering to ideal requirements is still a challenging task.

Among all the materials and manufacturing methods for SERS substrates [153], ion-exchange in glass may represent an effective route towards the realisation of performing and reliable SERS platforms thanks to the following features: (i) efficient SERS activity of these substrates (from 10^5 to 10^7 signal enhancement); (ii) intrinsically low material and processing costs; (iii) adequate sample reproducibility by opportunely tailoring the process and post-process parameters; (iv) long-term stability for the metal nanoparticles by their embedding within the glass matrix (typically, tens/hundreds nanometres under the glass surface) during a suitable post-process treatment; (v) actual sample reusability by partial exposure of noble metal nanostructures at the surface and their further refresh through suitable chemical and/or thermal steps [105,154].

3.2.1. Glasses for SERS Substrate by Ion-Exchange Technique

Soda-lime and float glasses with their large commercial availability, high transparency, good chemical durability, high sodium content, and ion inter-diffusion coefficients are the most popular material for the realisation of SERS substrates by ion-exchange process. The presence in trace of some impurities, such as iron or arsenic, within these glasses promotes the reduction of the exchanged noble metal ions into metallic clusters during the subsequent sample treatment [100,155–158].

Not to be forgotten are the commercial borosilicate glasses (e.g., Schott BK7 or Corning 0211 glasses) which, with their higher purity, are the most common substrates for the realisation of integrated optical devices by ion-exchange process. Although the almost total absence of impurities does not make them suitable for the formation of metal nanoparticles by chemical reduction processes, recent studies have shown the possibility of obtaining efficient SERS substrates also by using these highly pure glasses [159].

Finally, it should also be mentioned that, together with these widely used glass systems, phosphate or fluorophosphate glasses are recently gaining ever-increasing interest due to their lower processing temperature, higher diffusion coefficient for noble metal ions, and better reducing properties thanks to the large concentration of non-bridging oxygens generated during the glass synthesis [99,160].

Having incorporated the metal ions in the glass matrix by the ion-exchange process, one can promote their precipitation into metal nanoparticles by adopting different strategies, as described below.

3.2.2. Thermal Annealing in Air (O_2 Atmosphere)

The most common and simple strategy is represented by a thermal annealing treatment, carried out in air (O_2) in a suitable furnace, at a temperature close to the transition temperature T_g of the glass. The mechanism that regulates the formation of silver NPs inside an ion-exchanged soda-lime glass by means of this thermal post-process is well explained in the work of A. Simo et al. and outlined in Figure 4 below [106].

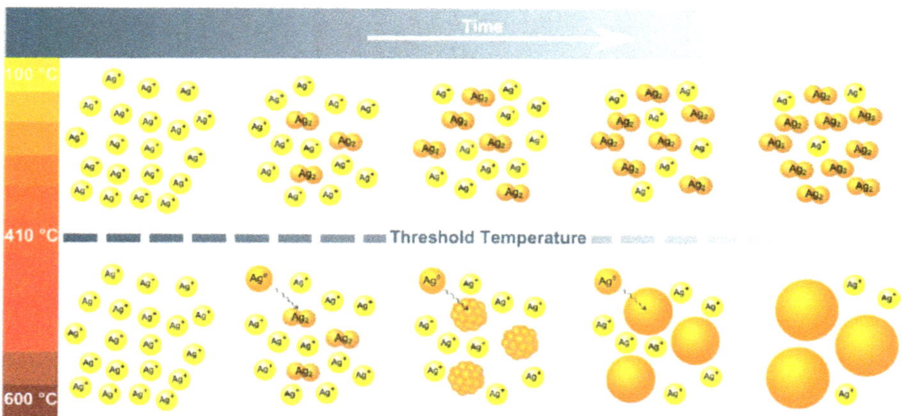

Figure 4. Outline of the mechanism regulating the formation of silver clusters and particle growth in soda-lime ion-exchanged glass under thermal annealing in air at different time and temperature. Reprinted with permission from [106]. Copyright © 2012, American Chemical Society.

After the ion-exchange process, only Ag+ ions are present within the glass. By increasing the annealing time and temperature (providing that the temperature remains below the critical threshold of about 400 °C), there is an increasingly large formation of silver clusters, which occur in the status of dimers Ag_2 (diameter size d < 1 nm). These small clusters are stable at elevated temperatures and remain within the glass matrix. The presence of these non-plasmonic nanostructures is the first responsible for the photoluminescence enhancement of the ion-exchanged sample. Thermal treatment performed beyond the aforementioned threshold temperature, which corresponds to the decomposition temperature of silver oxide, leads to the formation and the further growth of metallic silver NPs as the exchange time increases and the process temperature approaches Tg. In fact, in this temperature range, the number of reducing agents—the non-bridging oxygens (NBOs)—increases inside the glass matrix while the number of silver clusters remains unchanged. Consequently, the existing clusters begin to grow through the progressive addition of silver atoms Ag0 (monomers) towards the formation of ever larger plasmonic metal NPs, while the overall concentration of the particles remains constant.

Therefore, for a given temperature, the final size of these metal nanostructures is determined by the concentration of the dimers Ag_2 and the duration of the heat treatment. Typically, the formation of these particles is limited to the layers immediately below the glass surface, at a depth of a few tens/hundreds of nanometres, due to temperature-induced diffusion effects. Differently from the silver cluster case, the presence of these plasmonic metal nanoparticles produces a quenching effect on the photoluminescence intensity of the ion-exchanged specimen, especially when their size reaches several tens of nanometres or, at most, a few hundred nanometres [106].

3.2.3. Thermal Annealing in H_2 Controlled Atmosphere

Another thermal post-processing strategy for a glass exchanged with noble metal ions involves the heating of the sample in a controlled hydrogen (H_2) atmosphere. Although this technique requires larger precautions for the operator safety in comparison with that performed in air (O_2), this thermal post-process favours the formation of both glass-metal nanocomposite (GMN) species within the glass and metal island film (MIF) on the glass surface due to the out-diffusion mechanism of the noble metal atoms as a result of the hydrogen penetration into the glass [161]. However, under particular conditions relating to hydrogen processing, it is possible to favour the formation of MIF species on the glass surface by minimizing that of GMN within the host material, as well reported in the works of A. Lipovskii and collaborators, to which we refer for more details [162–164]. In brief, the authors employed a thermal poling technique, through the use of a suitably shaped anodic electrode applied to the ion-exchanged silicate glass sample, in order to modify the distribution of noble metal ions (e.g., Ag^+) and thus control the formation of the metal nanoislands by means of a soft thermal annealing step in a hydrogen atmosphere. Figure 5 explains this approach together with an atomic force microscopy (AFM) image of the nanoislands so formed. Depending on the electrode periodicity, arrays of nanoislands can be obtained on the surface of the glass sample.

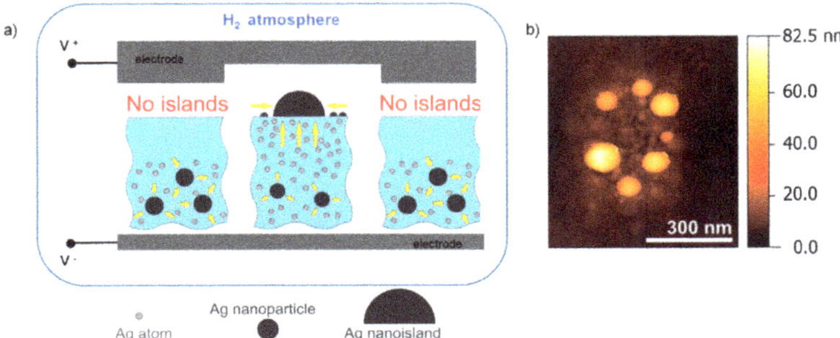

Figure 5. (**a**) A sketch of the mechanism of silver nanoislands growth by poling treatment following by a thermal annealing treatment in H_2 controlled atmosphere; (**b**) an AFM image of a nanoisland group in a not poled region after annealing treatment in H_2. Figure adapted from paper [164] under the terms of the Creative Commons Attribution 4.0 International License.

During the poling step, the simultaneous presence of electrostatic repulsion effects generated by the applied DC electric field and those linked to the process temperature favours the migration of the silver ions towards deeper layers of the glass in correspondence of the poled region. Nevertheless, the low poling temperature (T_p ~250 °C) and the lack of impurities in the pristine silicate glass do not favour the formation of silver atoms and metal nanoparticles in this area. Therefore, if the process parameters (i.e., duration and temperature) of the following thermal annealing in hydrogen are selected in such a way that the hydrogen penetration depth is not enough to reach the diffused silver ions, then any possibility of their reduction is precluded, and the poled zone remains almost empty of silver atoms and nanoparticles. On the contrary, in correspondence with the non-poled zone, the hydrogen in-diffusion is accompanied by an out-diffusion of the silver ions with their consequent reduction and formation of metal nanoislands on the glass surface [163]. A similar result was found by E. Kolobkva et al. in fluorophosphate glasses [99]. The possibility of drawing 2D-patterned nanoisland structures on the glass surface makes these substrates ready for their use in SERS applications.

3.2.4. Laser Irradiation Techniques

Besides the aforementioned thermal post-processes, other annealing techniques involve the use of continuous [102,165–167] and/or pulsed lasers [98,101,168–171] working at different wavelengths of the spectrum (from X-ray to ultraviolet (UV) and visible (VIS) regions up to infrared (IR) ones) and able to effectively reduce the noble metal ions in metal NPs. In this case, the mechanism of nanoparticle formation depends on the power of the laser source at the surface of the ion-exchanged specimen. In fact, in many cases the interaction between the laser beam and the sample is of thermal nature, as occurs for silver-ion-doped glasses when the working wavelength of the light source falls in the sample absorption region (i.e., from the UV to visible band, depending on the electronic status of silver). If the power level of the laser is relatively low, so as to ensure that the temperature of the irradiated area remains close to the glass transition temperature T_g during the whole process, then the reduction in silver atoms follows a trend almost similar to that of the thermal annealing mentioned above (in O_2 environment): laser irradiation produces free electrons from the non-bridging oxygens (NBOs) which are captured by the Ag^+ ions to form neutral Ag^0 atoms (metal silver atoms), starting their following growth process in metal NPs. As in the previous case, the nanoparticles thus formed are completely embedded within the glass matrix and a further etching step is required to make them partially exposed on the surface. This makes them available for SERS applications.

Conversely, at higher laser power levels, the temperature of the glass surface becomes greater than the glass T_g and the sample, in the irradiated region, behaves as a molten glass. In this liquid state, some of the silver nanoparticles are formed in a deeper region of the irradiated area while others may migrate to the glass surface where they are constrained to remain under the effect of the surface tension. Here, following the temperature distribution, they move towards the edges of the laser beam where they continue to aggregate forming larger structures. Once the laser is turned off, these nanoparticles remain frozen and partially exposed to the air [102,165,167]. In this way, these metal nanostructures are ready for SERS sensing and no further etching step is required. If compared with the standard thermal annealing process in air, the one induced by laser irradiation is faster in terms of nanoparticle formation and offers the non-negligible advantage of being able to draw ordered structures of nanoparticles—equally spaced from each other—so that the SERS response of the substrate is enhanced [102,166]. Figure 6 sketches the mechanism of metal nanostructure formation by high power laser beam in the case of silver ion-exchanged glass (Figure 6a). Figure 6b reports a SEM image of the edge region between irradiated and not-irradiated area. The lack of visible nanostructures on the surface of the exposed region suggests the possibility to obtain silver nanoparticles buried beneath the glass surface. Conversely, densely packed spherical silver NPs, whose size distribution ranges from a few to several tens of nanometres, are clearly visible on the ion-exchanged glass surface at the border of the irradiated area. Figure 6c shows a SEM image of a quasi-periodic arrangement of silver NPs obtained with this technique. Finally, Figure 6d represents a zoom image of an array of silver nanoparticles. The alignment of the nanostructures appears quite regular.

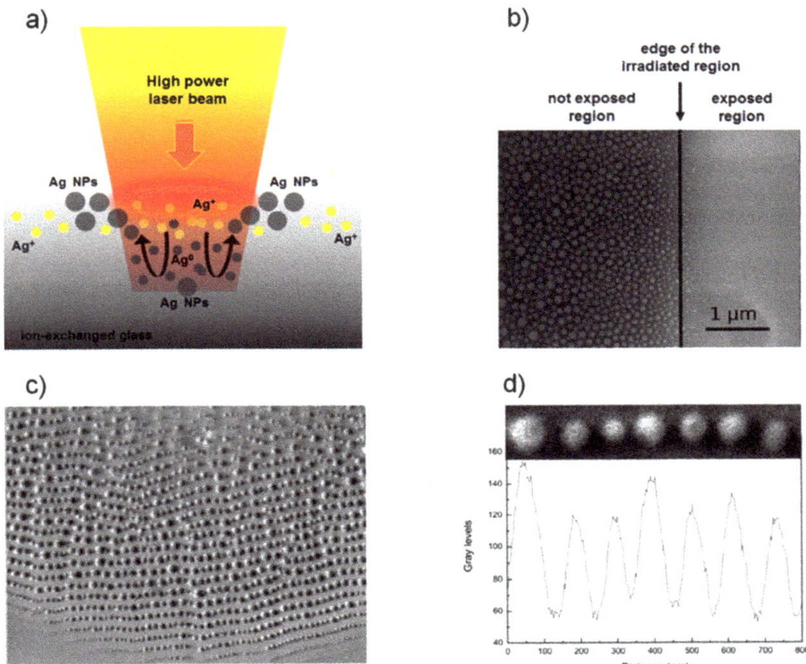

Figure 6. (**a**) Growth process of metal silver nanoparticles on the ion-exchanged sample surface at the edge of an exposed zone by high power laser irradiation; (**b**) a comparison between the distribution of silver nanoparticles in the exposed and not-exposed region, respectively (adapted with permission from [165] © The Optical Society); (**c**) a SEM image of a quasi-periodic arrangement between silver NPs; (**d**) an ordered silver nanoparticle array, with its surface profile plot, obtained by laser beam irradiation technique (reprinted with permission from [166] © The Optical Society).

Obviously, once exposed, these silver NPs can undergo oxidation effects that affect their performance in terms of SERS response. A possible solution to the problem could be to carry out a low temperature thermal annealing in air ($T_{ann} \sim 200\,°\,C$) in order to reduce to metallic state the thin layer of silver oxide (Ag_2O) thus formed on the nanoparticles surface, which is mainly responsible for the deterioration of the SERS enhancement [102]. Another approach to prevent the aging of silver nanoparticles concerns the deposition of a thin film of transparent dielectric material (e.g., SiO_2, TiO_2, Si_3N_4, Al_2O_3, etc.), provided that the thickness of the deposited layer is such as not to excessively weaken the plasmon resonance and the consequent field enhancement effect. For this purpose, tight thickness control deposition techniques, such as atomic layer deposition (ALD) and plasma-enhanced chemical vapour deposition (PECVD), are preferable [172,173].

3.2.5. Metal Thin Film Deposition for Noble Metal Ions Reduction in Ion Exchange Process

The possibility of realizing SERS active metal nanostructures on high quality, free from impurities, commercial glasses (such as, Corning 0211 borosilicate glass) by means of ion-exchange process can be favoured by applying on the sample surface a metal thin film working as a mask or electrode. In both cases, the presence of the film plays the role of reducing element for the silver ions introduced into the glass following the ion-exchange, as well investigated in the works of Y. Chen and co-workers [103,104,159].

In the masked ion-exchange process in silver molten salts, the reduction mechanism of silver ions in metallic form is strictly related to the presence of electrical potential differences among the molten salt, the glass and the metal mask. This produces an ionic current flow: the silver ions migrate from the molten salt into the glass through the mask openings, concentrating in the region under the mask edges. Here, by capturing an electron made available by the metal mask, the silver ions are progressively reduced to metallic state [103,104]. Figure 7 shows the effect of a metal mask on the reduction of silver ions following the ion-exchange process.

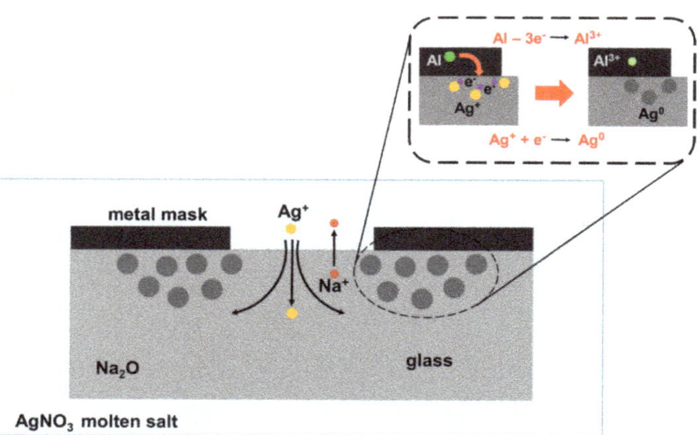

Figure 7. Deposition of silver nanoparticles under the mask edge by masked ion-exchange process (cross-sectional view). In inset, the reduction mechanism of silver ions in metal nanoparticles under an aluminium mask.

The main advantage of this photolithographic approach is to provide the formation of well-defined patterns of silver nanoparticles, aligned with ion-exchange channel waveguides, without any thermal post-processing of the sample. Furthermore, this method represents a valid alternative to that of laser irradiation to realize integrated sensors, based on optical waveguides and SERS.

In the second approach, the deposed metal thin film works as an electrode in a double ion-exchange process. The glass sample is loaded with silver ions through a simple ion-exchange in molten salt. Subsequently, a thin metal film of aluminium is deposited on one of the two sample faces. The sample is then immersed in a potassium molten salt bath. In this way, the overall system represented by the metal thin film, the ion-exchanged glass and the molten salt acts as a real galvanic cell. In fact, two electric potentials are soon established on the two faces of the glass. The first, positive type, is located near the glass-salt interface and takes place from the different inter-diffusivity of the ionic species involved in the process. In this region, the Ag^+ and Na^+ ions diffuse faster in the molten salt than the K^+ ions in the glass. The second, negative type, is generated near the other thin film-molten-salt interface. Here the thin aluminium film tends to release electrons to the Ag^+ ions in the glass while Al^{3+} ions migrate to the molten potassium salt. This electrical potential difference produces an ionic current inside the glass which forces the silver ions to move to the surface of the specimen, in an area immediately below the metal electrode. In this region, the silver ions are reduced in metallic form through the acquisition of an electron coming from the metallic film. The formation of silver nanoparticles is also promoted by the continuous flow of silver ions coming from the deeper regions of the glass towards the interface with the aluminium thin film [159]. Figure 8 describes the mechanism mentioned above (Figure 8a), reporting a scanning electron microscope (SEM) cross-section image of the silver nanoparticles distribution under the sample surface (Figure 8b).

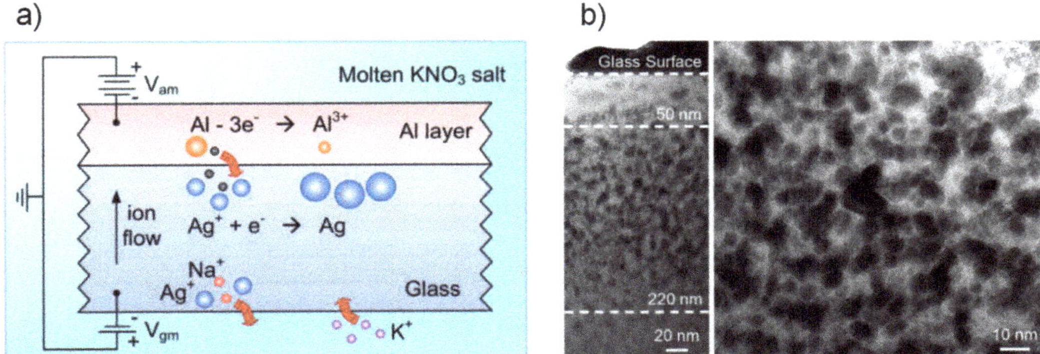

Figure 8. (**a**) Mechanism for the formation of silver nanoparticles in a double ion-exchange using an aluminium thin film as electrode; (**b**) distribution of silver nanoparticles, under the sample surface, following the process (SEM image of the sample cross section near to the surface). Reprinted with permission from [159] © The Optical Society.

The main feature of this method is to offer a simple and low-cost solution for the production of homogeneous and large-scale SERS substrates.

In both approaches described above, the formed silver nanoparticles are embedded within the glass. Therefore, a subsequent soft etching step in HF acid is necessary to expose them at the surface level and make these substrates functional for SERS applications.

We would also like to mention the work of P.L. Inacio et al. for its simplicity. Their investigation, using optical and AFM (atomic force microscopy) measurements, showed a migration of silver Ag^+ cations from inside of the waveguide to its surface at room temperature. Once exposed to air at room temperature, the cations undergo an oxidation process changing the SERS response of the substrate [174]. The advantage of this approach is that it does not require any additional process other than that represented by the aging over time of the exchanged sample. On the other hand, the main drawbacks lie in the rather long aging times, the instability of the SERS substrate over time, and the poor adhesion of the metal nanoparticles formed on the glass surface, which makes them easily removable under the minimum mechanical action or event.

SERS platforms created by ion-exchange are still in their early state, since the main scientific contributions reported so far in the literature demonstrate only proof of concepts of their operation through the use of aromatic molecules (e.g., rhodamine, etc.). However, thanks to their interesting performances both in terms of signal enhancement and detection limit [154], it is likely that soon these substrates will actually be used for the detection of biological targets, such as DNA sequences, proteins, bacteria, etc., as is happening for SERS substrates obtained with different technological approaches [97,151].

4. Perspectives & Future Insights

The high versatility of the ion-exchange process in glass finds new and fascinating perspectives of application in the field of photonics, some of them in progress while others still deserve to be investigated.

4.1. All-Glass Flexible Photonics

As happened for electronics, flexibility is also becoming a fundamental paradigm in photonics, favouring the integration of the respective devices in flexible and stretchable substrates and contributing to expand their application scope, from the field of optical interconnections to that of new systems for the continuous monitoring of patient health, or that of implantable devices for minimally invasive medicine, up to the development of a new class of optical sensors and sources [175–179].

Polymers are certainly the main substrates for the development of this novel technology thanks to their intrinsic mechanical flexibility. In order to give some examples, it is worth listing the possibility of making silicon photonic devices on plastic substrates by means of low temperature approaches (e.g., transfer-and-bond technique, direct fabrication process flow) [180,181]; the deposition of glass-based 1D photonic crystals and erbium activated planar waveguides by radio frequency (RF) sputtering on flexible polymeric substrates [182]; the realisation of stretchable platforms in which single-mode photonic devices, made from high-index chalcogenide glasses and low-index epoxy polymers, are monolithically integrated on polydimethylsiloxane (PDMS) elastomer substrate. In the latter case, local substrate stiffening and spiral based waveguides provide protection to functional photonic components and interconnections with very low radiative optical losses, respectively [183,184]. Polymeric materials, however, exhibit various drawbacks such as a limited range of refractive indices, not negligible optical absorption in the near-infrared region, low process temperature and poor chemical stability. All these factors generally represent serious side effects against their integration and processing with other materials.

Conversely, glass, with a wide range of refractive indices, high transparency from visible to infrared, excellent chemical and thermal stability, as well as low permeation rates for ambient gases and water, may represent a valid alternative as flexible platforms for the realisation of photonic devices provided that their brittleness is somehow mitigated [185]. In this context, J. Lapointe et al. demonstrated the possibility to obtain single mode, low-loss (<0.15 dB/cm), laser-written channel waveguides in flexible chalcogenide glass tape [186]. Nevertheless, among all the glass formulations that can be used for the purpose of flexible photonics, borosilicate glass represents the most widely used material due to its intrinsic characteristics of mechanical strength. In particular, alkali-free commercial borosilicate glasses with extremely thin thicknesses (<100 µm) are currently made available by major glass producers in the format of sheets or rolls. This is the case of Willow® glass from Corning and AF32® from Schott just to mention a few [187,188]. The flexibility of these glasses is conferred through particular manufacturing processes—such as down-draw or roll-to-roll (R2R) techniques—capable of guaranteeing ultra-thin thicknesses to the finished product with an extremely high surface quality, similar to that of a pristine glass [189,190].

As a white sheet of paper waiting to receive the characters from the pen of an inspired writer, so Willow glass with different thickness can be used as a substrate for the realisation of photonic circuits as demonstrated by S. Huang and co-workers. In this case, the writing instrument is represented by a femtosecond laser capable of locally modifying, in the irradiated area, the refractive index of the pristine glass [191]. Figure 9 shows photos of bent waveguides fabricated in Willow glass with different thickness. The scattered green light, coming from a laser source working at 532 nm and injected inside the guiding structures by a butt-coupled optical fibre, visually highlights the actual presence of these channel waveguides.

By optimising the laser processing parameters in terms of speed and energy pulse, it is possible to obtain single-mode channel waveguides at the working wavelength of 1550 nm with propagation loss as low as 0.11 dB/cm in all flexible Willow glass formulations. In particular, for the selected refractive index change ($\Delta n \sim 5 \times 10^{-3}$), the bending losses of the waveguides are practically negligible for a curvature radius ≥ 1.5 cm. Furthermore, the high thermal stability of these optical devices (up to 400 °C), together with the possibility of writing Bragg grating waveguides, opens up new potential scenarios for the realisation of more complex photonic circuits on these commercial flexible glasses. Due to their alkaline-free content, these glasses find one of their main uses in the development of flat panel displays for high-tech applications [192].

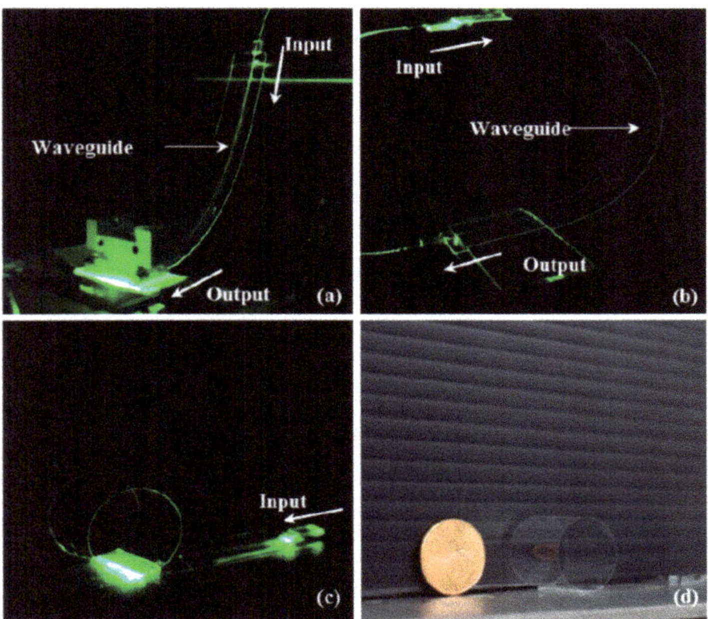

Figure 9. Photos of bent waveguides written by femtosecond laser in Willow glass strips with different thickness and bending radius (**a**) 100 μm thickness and bending radius 13.5 cm; (**b**) 50 μm thickness and bending radius 2.1 cm; (**c**) 25 μm thickness and bending radius 1.0 cm; (**d**) the same sample reported in (**c**) compared with the size of a small coin. Reprinted with permission from [191] © The Optical Society.

The possibility of further strengthening these ultra-thin glasses could prevent the formation of any surface flaws making the customary handling and processing (e.g., glass cleaning, cutting, and lamination) procedures of these products less critical. Nevertheless, thermal tempering does not represent a suitable solution for strengthening these display glasses due to their thinness and low thermal expansion coefficient. On the other hand, the total absence of alkaline ions (e.g., sodium ions, Na^+) in these products would seem to preclude the use of any chemical treatment aimed at further strengthening them. In this context, the interesting work by Mauro et al. would seem to offer an elegant solution to the problem, bypassing this impasse through the use of a particular ion-exchange process in alkali-free boroaluminosilicate glass [193]. The idea behind this original method is to replace ions such as Ca^{2+} and B^{3+} present inside the pristine glass matrix with Ba^{2+} ions coming from an ion source where the sample is immersed. Unfortunately, the kinetics of this kind of ion-exchange would be decidedly limited due to the considerable difference in inter-diffusivity between the two ionic species involved in the process, the alkaline and alkali ones, respectively. Moreover, these alkaline salts present high melting point and chemical corrosiveness that could irreversibly damage the glass surface during the exchange process, making their use as ion sources in a standard ion-exchange practically unfeasible. For all these reasons, the new chemical strengthening treatment is carried out in aqueous solutions of alkaline salts at a relative low temperature and under pressure in order to keep the water stable in its liquid state [194]. Proceeding in this way, the authors demonstrated the possibility to reach surface compressive stresses (CS) of the order of 200 MPa and a depth of layer (DoL) of only a few microns (i.e.,: <2 μm for barium nitrate). This technique can be applied to a wide range of alkali-free glasses, such as ultra-thin

Willow glass, contributing to improve mechanical performance and thus helping to make all-glass flexible photonics more feasible.

A second promising approach towards the development of all-glass flexible photonics is the one that involves the use of alkali-based glass formulations (i.e., sodium aluminosilicate glass). In fact, recent analytical studies conducted by G. Macrelli and co-workers shed new light on the effects induced by the bending stresses on both the mechanical strength and optical properties (e.g., in terms of refractive index) in the two main silicate glass formulations considered here, the alkali-free and the alkali-based ones [195,196]. In general, the thinner the glass, the lower the surface stress resulting from a bending action. This allows to reach ever smaller bending radii, within the limits given by the mechanical strength of the material related to the presence of surface flaws. When subjected to extreme curvature, the glass develops high tension stresses that can lead either to its immediate breakage or to the progressive growth over time of micro-cracks, generated both by the effect of the continuous applied tension and by the action of water vapour on surface flaws, resulting in consequent definitive damage of the material. To overcome this problem, the strategy based on alkali-free glasses dictates a complete sealing of their surface, to make it resistant to the deteriorating action of water vapours on any surface flaws. On the other hand, the strategy based on alkali-silicate glasses involves the use of ion-exchange, capable of strengthening the glass by means of a residual tension profile induced on the surface of the sample following the same process. This suggests that the achievement of the surface flaws stability in extreme bending conditions for the glass represents a key factor to be considered.

From the mechanical analysis carried out on the two types of samples mentioned above it emerges that, under the same extreme bending conditions, the alkali-based silicate glasses strengthened by ion-exchange generally allow larger thickness, overcoming the limitations imposed on not chemically treated alkali-free or alkali-based glasses. Moreover, from an optical point of view, the response of the two types of glass formulation under the same bending stress appears very different in terms of refractive index. In particular, for glasses strengthened by ion-exchange, the presence of residual surface tensions is associated to a local volume change of the material. This induces a further birefringence phenomenon that well overcomes that coming from the bending action of the glass sample. As a result, a strong refractive index gradient occurs at the surface of these exchanged ultra-thin glasses. This effect must be taken into account when designing all-glass photonic devices on this kind of flexible substrates.

From what has been reported up to now it is clear the key role that ion-exchange, together with the glass material engineering, can play in the creation of increasingly performing flexible substrates towards the achievement of an all-glass flexible photonics as valid alternative to that developed on polymeric materials [178].

4.2. New Insight towards the Development of Optical Devices Loaded with Noble Metal Ions/Nanoparticles by Ion-Exchange

Another intriguing prospect of the ion-exchange process that deserves further investigation in the near future is linked to the possibility of realizing new low-cost photonic systems with potentially improved characteristics for a wide application range.

Let consider, for instance, the case of whispering gallery mode (WGM) optical microcavities, which continue to gain a growing interest in the field of photonics for their unique and extraordinary properties. In these microstructures, the light, suitably injected into the cavity, propagates grazing the surface of the resonator (few microns in depth) through subsequent total internal reflections along an equatorial plane. Due to their low material losses and high surface quality, these optical microcavities present an extremely high value of their quality factor Q defined as the ratio between the wavelength λ at which a resonance occurs and the linewidth $\Delta\lambda$ of the resonant wavelength. Moreover, their small mode volumes and long storage lifetimes for the photons guarantee high energy density levels and very narrow linewidths for the resonant modes sustained by these resonant microstructures. All these properties make such optical devices of great interest for applications in lasing, non-linear optics, optical communications and sensing [197–199].

If fused silica represents one of the basic materials for the realisation of WGM microresonators with extremely high Q factor thanks to its excellent chemical durability and transparency in a wide range of wavelengths [200], soft glasses are now gaining an increasing interest as a bulk material for these WGM microcavities due to their better versatility in terms of material properties, low phonon energy, and high rare earth solubility [201–205].

Based on this promising background in terms of materials for WGM microresonators, the basic idea could be to apply ion-exchange process to load the soft glass material constituting these microresonators with noble metal ions and/or nanoparticles in order to create hybrid devices, able to make the most of the intrinsic characteristics of the WGMs joined with those of the ion-exchanged material.

For example, let's consider the case of a lasing emission by up-conversion phenomenon in WGM glass microspheres doped with rare earth ions [206–210]. The possibility to nucleate and grow noble metal nanostructures (i.e.: in the form of dimer clusters and/or metal nanoparticles with few nanometre sizes) within the active medium of the cavity following ion-exchange and thermal post-process may enhance the luminescent efficiency of the rare earth element and its emission by surface plasmon resonance (SPR) and/or energy transfer (ET) mechanisms, as observed for the planar substrates [211–213]. Moreover, an increase in the up-conversion signal comes from the same WGM microcavity which, supporting high recirculating light energy densities, allows the photons to interact several times both with the rare earth ions and metal nanostructures generally located at a distance of few tens/hundreds of nanometres below the cavity surface within the WGMs action field. In particular, if these nanostructures are represented by small plasmonic NPs (e.g., few nanometres in size), then a combination between the two resonant phenomena—the one linked to the host WGM microcavity and the plasmonic one of the nanoparticles—may occur generating a further significant signal enhancement.

Obviously, the ion-exchange process with the consequent presence of noble metal nanostructures may increase the losses in the cavity reducing its Q factor value. In this regard, it should be noted that the thermal annealing process in air may offer a double advantage: formation of noble metal nanoparticles embedded in the glass and smoothening of the WGM microcavity surface as a result of temperature and surface tension. As proof of this statement, K. Milenko and co-workers recently demonstrated how the quality factor Q of a silver iodide phosphate glass microsphere remains almost unchanged ($\sim10^4$) before and after the formation of silver nanoparticles following a thermal annealing treatment in air. The size of silver nanoclusters so obtained was around 350 nm [214]. This encouraging result suggests that the achievement of hybrid WGM microcavities, both in passive and in active glasses, by means of the ion-exchange technique may not be such a distant possibility.

Currently, one of the most promising glass formulations for the realisation of active WGM microresonators is represented by phosphate glasses able to guarantee high Q factor values, above 10^6, for these optical devices both in bulk and coated configuration [201,215]. Hence, these host materials may represent good candidates for the realisation of hybrid active WGM microresonators by ion-exchange process. In this context, it will be mandatory to tailor both the ion-exchange parameters and those of the subsequent post-processing treatment to optimize the size of the noble metal nanostructures (e.g., few nanometres in diameter), minimizing losses in the cavity and keeping the quality factor Q for the resulting hybrid active WGM microcavities as high as possible.

The considerations made above for bulk microspheres continue to be valid also for coated ones such as, for instance, a silica microsphere with a sol-gel thin film coating of compound glass. Applications in the development of visible laser sources for displays should be expected.

The availability of noble metal nanoparticles embedded in a WGM cavity of compound glass could also favour non-linear optical response of the medium that should be further increased by the presence of plasmonic effects. For instance, silver NPs with their large third order nonlinear properties and optical absorption transitions with high oscillator strengths are the main candidates for the development of all-optical signal processing devices such as efficient optical limiters, which require high nonlinear refractive index n_2 materials [216–218]. Also in this case, the general considerations made previously in terms of the optimisation of both process parameters and quality factor Q are still valid.

It is important to note that, although the plasmonic absorption peak is generally located in the visible region of the spectrum, the absorption band of glasses with embedded silver NPs presents a tail extending to the near infrared region. This suggests how the local field effect induced by SPR phenomenon would have an impact also in this region of the spectrum [159,219]. This would allow to enhance the visible up-conversion emission in erbium doped glasses with embedded silver NPs. The enhancement in the up-conversion intensity may be also attributed to the local field effect induced by the presence of silver NPs [220].

Figure 10 shows a potential up-conversion emission enhancement in the particular case of an Er^{3+}-doped WGM microsphere with embedded silver plasmonic small nanoparticles.

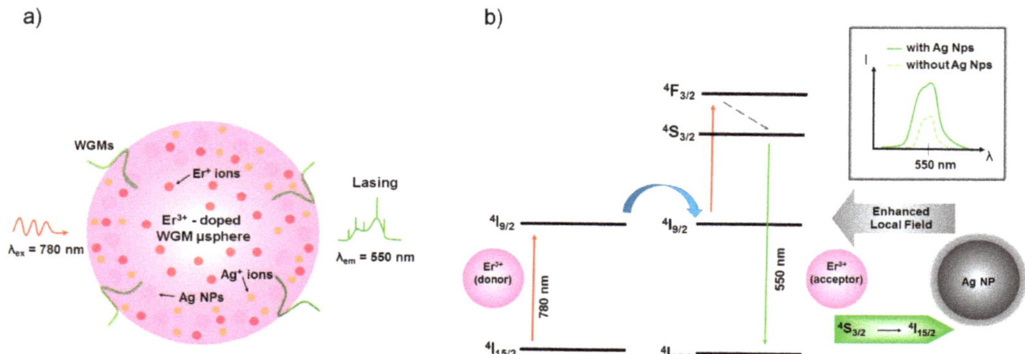

Figure 10. (**a**) Scheme of the up-conversion lasing emission in a Er^{3+}-doped WGM microsphere with embedded silver nanoparticles. (**b**) Potential emission enhancement process in presence of Ag NPs. In the inset, the expected qualitative trend of the erbium up-conversion signal with (continuous line) or without (dashed line) plasmonic silver nanoparticles.

It should also be mentioned here that a further important contribution for the achievement of hybrid WGM microcavities should be offered by the research effort towards the development of novel and efficient glass formulations, suitable for ion-exchange process, with the aim to reduce material absorption losses and scattering effects.

Another noteworthy application of the ion-exchange technique joined with glass WGM microcavity could be represented by growing noble metal NPs (e.g., metal nanoislands or an array of few ordered and equidistant nanoparticles, working as hot spots and based on localized surface plasmon resonance (LSPR) phenomenon), directly on the surface of an ion-exchanged microresonator—in correspondence with its equatorial plane—by means of a thermal post-process performed with laser irradiation. This methodology could represent a valid alternative to the widely used one, which implies the chemical immobilisation of plasmonic nanoparticles on the surface of the resonant microcavity. In comparison to this technique, the strategy based on laser irradiation reduces the processing times and allows to have nanoparticles partially exposed on the surface, therefore more resistant to the various washing steps generally required in biosensing applications. Figure 11 shows the growth mechanism of plasmonic silver nanoparticles on the surface of an ion-exchanged WGM microsphere by laser beam irradiation technique.

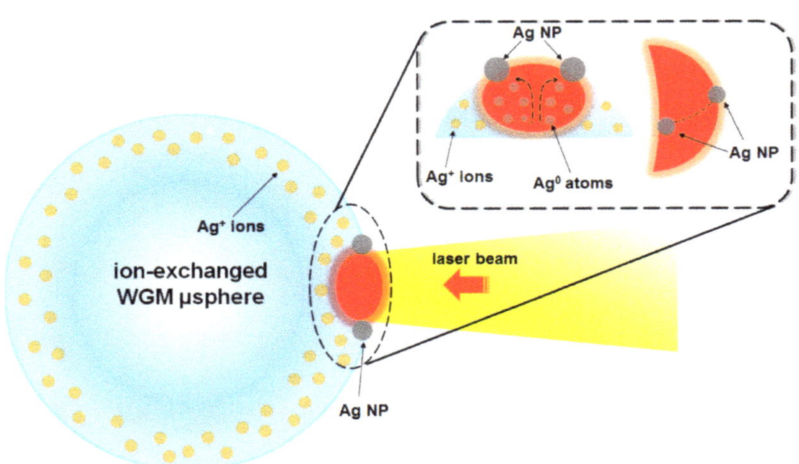

Figure 11. A sketch of the formation mechanism of silver nanoparticles on the WGM microsphere surface by means of a laser beam irradiation. The nanoparticles (NPs) result partially exposed as reported in detail in the inset.

Another strategy to partially expose the noble metal NPs on the surface of a WGM resonant glass microcavity could be represented by a chemical etching step in an extremely dilute solution of hydrofluoric acid (HF), in order to preserve as much as possible a low value of surface roughness for the microresonator.

In recent years, with the improvement of manufacturing techniques, there has been a continuous and growing interest in the realisation of new hybrid optical fibres (HOFs), in which the co-presence within a silica host fibre of multi-materials (e.g., glasses, crystals, semiconductors, etc.) and different guiding structures (e.g., non-linear waveguides, photonic crystals, semiconductor junction, optoelectronic elements, etc.) contributes to broaden their application horizon towards the achievement of an all-fibre photonic platform [221–226]. In this context, ion-exchange in molten salts of noble metals with subsequent thermal post-process may induce the nucleation and growth of plasmon nanoparticles at the end of a possible HOF, whose core is made up of a compound glass suitable for ion-exchange (e.g., silicate, phosphate, tellurite, etc.). Alternatively, the end of a multimode silica optical fibre loaded with a thin layer of soft glass could be used for the same purpose. In the near future, these strategies could represent an effective low-cost solution towards the realisation of optical fibre probes for SERS applications in comparison with current lithographic techniques, such as electron beam lithography (EBL) or nanoimprinting lithography (NIL), surely more effective but definitely complicated and expensive technologies [227,228]. Figure 12 represents some configurations for the development of novel fibre probes for SERS applications. In particular, Figure 12a reports the concept of multimaterial HOF while Figure 12b,c show two possible schemes, one based on HOF and the other on a multimode optical fibre (MOF).

Finally, another application that deserves to be investigated concerns the fabrication of compound glass nanotips [229] with embedded noble metal nanoparticles—following ion-exchange and thermal annealing processes—for SERS and tip enhanced Raman spectroscopy (TERS) applications [230,231].

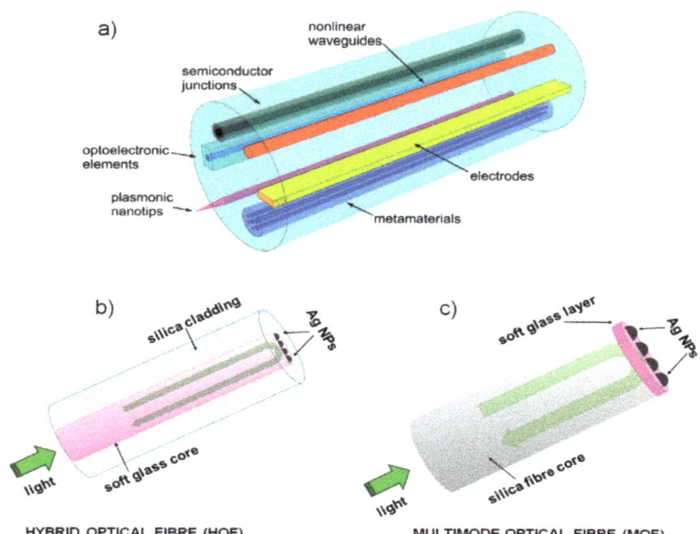

Figure 12. (**a**) The concept of multimaterial hybrid optical fibre (HOF) (figure reprinted from paper [222] under the terms of the Creative Commons Attribution 4.0 International License); (**b**) a possible scheme for a SERS fibre probe based on HOF; (**c**) a possible scheme for SERS fibre probe based on multimode optical fibre (MOF).

5. Conclusions

In this review paper, we have shown how the glass and ion-exchange pair, far from being an obsolete technology, still plays a key role in various technological fields with interesting applications and industrial developments that also have repercussions at the level of everyday life. Moreover, this ancient and indissoluble synergy between material and manufacturing process may lead in the near future to the realisation of new and intriguing photonic devices with further interesting applications in optical communications, non-linear optics, and sensing.

Author Contributions: Conceptualization, S.B.; writing—original draft preparation, review and editing, S.B. and S.P.; supervision, G.C.R. All authors have read and agreed to the published version of the manuscript.

Funding: This research was partially funded by the European Community and the Tuscany Region (Italy) within the framework of the SAFE WATER project (Horizon 2020 Research&Innovation program and the ERA-NET "PhotonicSensing" cofund -G.A.No 688735).

Acknowledgments: The authors wish to acknowledge the crucial and pleasant role played by discussions and exchange of opinions and information with their colleagues (in random order) Gualtiero Nunzi Conti, Silvia Soria, Franco Cosi, Massimo Brenci, Daniele Farnesi, Riccardo Saracini, Maurizio Ferrari, Alessandro Chiasera.

Conflicts of Interest: The authors declare no conflict of interest.

References

1. Rehren, T.; Freestone, I.C. Ancient glass: From kaleidoscope to crystal ball. *J. Archeol. Sci.* **2015**, *56*, 233–241. [CrossRef]
2. Scott, R.B.; Brems, D. The archaeometry of ancient glassmaking: Reconstructing ancient technology and the trade of raw materials. *Perspective* **2014**, *2*, 224–238. [CrossRef]
3. Righini, G.C.; Laybourn, P.J.R. Integrated Optics. In *Perspectives in Optoelectronics*, 1st ed.; Jha, S.S., Ed.; World Scientific Publishing Co. Pte. Ltd.: Singapore; Hackensack, NJ, USA; London, UK; Hong Kong, China, 1995; Chapter 12, pp. 679–736.

4. Righini, G.C.; Pelli, S. Ion-exchange in glass: A mature technology for photonics devices. In Proceedings of the International Symposium on Optical Science and Technology, San Diego, CA, USA, 29 July–3 August 2001; Volume 4453, pp. 93–99. [CrossRef]
5. Pradell, T.; Molera, J.; Roque, J.; Vendrell-Saz, M.; Smith, A.D.; Pantos, E.; Crespo, D. Ionic-exchange mechanism in the formation of medieval luster decorations. *J. Am. Ceram. Soc.* **2005**, *88*, 1281–1289. [CrossRef]
6. Mazzoldi, P.; Sada, C. A trip in the history and evolution of ion-exchange process. *Mat. Sci. Eng. B* **2008**, *149*, 112–117. [CrossRef]
7. Mazzoldi, P.; Carturan, S.; Quaranta, A.; Sada, C.; Sglavo, V.M. Ion-exchange process: History, evolution and applications. *Riv. Nuovo Cim.* **2013**, *36*, 397–460. [CrossRef]
8. Schulze, G. Versuche über die diffusion von silber in glas. *Ann. Phys.* **1913**, *345*, 335–367. [CrossRef]
9. Kistler, S.S. Stresses in glass produced by nonuniform exchange of monovalent Ions. *J. Am. Ceram. Soc.* **1962**, *45*, 59–68. [CrossRef]
10. Acloque, P.; Tochon, J. Measurement of Mechanical Resistance of Glass after Reinforcement. In *Colloquium on Mechanical Strength of Glass and Ways of Improving It*; Union Scientifique Continentale du Verre: Charleroi, Belgium, 1962; Volume 1044, pp. 687–704.
11. Donald, I.W. Methods for improving the mechanical properties of oxide glasses. *J. Mater. Sci.* **1989**, *24*, 4177–4208. [CrossRef]
12. Gy, R. Ion-exchange for glass strengthening. *Mater. Sci. Eng. B* **2008**, *149*, 159–165. [CrossRef]
13. Karlsson, S.; Jonson, B. The technology of chemical glass strengthening—A review. *Glass Technol. Eur. J. Glass Sci. Technol. A* **2010**, *51*, 41–54.
14. Agrawal, G.P. Optical Communication: Its History and Recent Progress. In *Optics in Our Time*; Al-Amri, M., El-Gomati, M., Zubairy, M., Eds.; Springer: Cham, Switzerland, 2016; Chapter 8; pp. 177–199. [CrossRef]
15. Maurer, R.D.; Schultz, P.C. Fused Silica Optical Waveguide. US Patent 365995, 2 May 1972.
16. Keck, D.B.; Schultz, P.C. Method of Producing Optical Waveguide Fibres. US Patent 3711262, 16 January 1973.
17. Miller, S.E. Integrated Optics: An Introduction. *Bell Syst. Tech. J.* **1969**, *48*, 2059–2069. [CrossRef]
18. Izawa, T.; Nakagome, H. Optical waveguide formed by electrically induced migration of ions in glass plates. *Appl. Phys. Lett.* **1972**, *21*, 584–586. [CrossRef]
19. Giallorenzi, T.G.; West, E.J.; Kirk, R.D.; Ginther, R.J.; Andrews, R.A. Optical waveguides formed by thermal migration of ions in glass. *Appl. Opt.* **1973**, *12*, 1240–1245. [CrossRef] [PubMed]
20. Righini, G.C.; Chiappini, A. Glass optical waveguides: A review of fabrication techniques. *Opt. Eng.* **2014**, *53*, 071819. [CrossRef]
21. Thervonen, A.; West, B.R.; Honkanen, S. Ion-exchanged glass waveguide technology: A review. *Opt. Eng.* **2011**, *50*, 071107. [CrossRef]
22. Ramaswamy, R.V.; Srivastava, R. Ion-exchanged glass waveguides: A review. *IEEE J. Light. Technol.* **1988**, *6*, 984–1000. [CrossRef]
23. Honkanen, S.; West, B.R.; Yliniemi, S.; Madasamy, P.; Morrell, M.; Auxier, J.; Schülzgen, A.; Peyghambarian, N.; Carriere, J.; Frantz, J.; et al. Recent advances in ion-exchanged glass waveguides and devices. *Phys. Chem. Glasses: Eur. J. Glass Sci. Technol. B* **2006**, *47*, 110–120.
24. Äyräs, P.; Nunzi Conti, G.; Honkanen, S.; Peyghambarian, N. Birefringence control for ion-exchanged channel glass waveguides. *Appl. Opt.* **1998**, *37*, 8400–8405. [CrossRef]
25. Asquini, R.; D'Angelo, J.; d'Alessandro, A. A Switchable Optical Add-Drop Multiplexer using Ion-Exchange Waveguides and a POLICRYPS Grating Overlayer. *Mol. Cryst. Liq. Cryst.* **2006**, *450*, 203–214. [CrossRef]
26. D'Alessandro, A.; Donisi, D.; De Sio, L.; Beccherelli, R.; Asquini, R.; Caputo, R.; Umeton, C. Tunable integrated optical filter made of a glass ion-exchanged waveguide and an electro-optic composite holographic grating. *Opt. Express* **2008**, *16*, 9254–9260. [CrossRef]
27. Montero-Orille, C.; Moreno, V.; Prieto-Blanco, X.; Mateo, E.F.; Ip, E.; Crespo, J.; Liñares, J. Ion-exchanged glass binary phase plates for mode-division multiplexing. *Appl. Opt.* **2013**, *52*, 2332–2339. [CrossRef]
28. Montero-Orille, C.; Prieto-Blanco, X.; González-Núñez, H.; Liñares, J. A Polygonal Model to Design and Fabricate Ion-Exchanged Diffraction Gratings. *Appl. Sci.* **2021**, *11*, 1500. [CrossRef]
29. Bradley, J.D.B.; Pollnau, M. Erbium-doped integrated waveguide amplifiers and lasers. *Laser Photonics Rev.* **2011**, *5*, 368–403. [CrossRef]
30. Righini, G.C.; Brenci, M.; Forastiere, M.A.; Pelli, S.; Ricci, R.; Nunzi Conti, G.; Peyghambarian, N.; Ferrari, M.; Montagna, M. Rare-earth-doped glasses and ion-exchanged integrated optical amplifiers and lasers. *Phylos. Mag.* **2002**, *82*, 721–734. [CrossRef]
31. Pelli, S.; Bettinelli, M.; Brenci, M.; Calzolai, R.; Chiasera, A.; Ferrari, M.; Nunzi Conti, G.; Speghini, A.; Zampedri, L.; Zheng, J.; et al. Erbium-doped silicate glasses for integrated optical amplifiers and lasers. *J. Non Cryst. Solids* **2004**, *345–346*, 372–376. [CrossRef]
32. Righini, G.C.; Arnaud, C.; Berneschi, S.; Bettinelli, M.; Brenci, M.; Chiasera, A.; Feron, P.; Ferrari, M.; Montagna, M.; Nunzi Conti, G.; et al. Integrated optical amplifiers and microspherical lasers based on erbium-doped oxide glasses. *Opt. Mater.* **2005**, *27*, 1711–1717. [CrossRef]
33. Berneschi, S.; Bettinelli, M.; Brenci, M.; Dall'Igna, R.; Nunzi Conti, G.; Pelli, S.; Profilo, B.; Sebastiani, S.; Speghini, A.; Righini, G.C. Optical and spectroscopic properties of soda-lime alumino silicate glasses doped with Er^{3+} and/or Yb^{3+}. *Opt. Mater.* **2006**, *28*, 1271–1275. [CrossRef]
34. Ondráček, F.; Jágerská, J.; Salavcová, L.; Míka, M.; Špirková, J.; Čtyroký, J. Er–Yb Waveguide Amplifiers in Novel Silicate Glasses. *IEEE J. Quantum Electron.* **2008**, *44*, 536–541. [CrossRef]
35. Laporta, P.; Taccheo, S.; Longhi, S.; Svelto, O.; Svelto, C. Erbium–ytterbium microlasers: Optical properties and lasing characteristics. *Opt. Mater.* **1999**, *11*, 269–288. [CrossRef]

36. Della Valle, G.; Taccheo, S.; Sorbello, G.; Cianci, E.; Foglietti, V.; Laporta, P. Compact high gain erbium-ytterbium doped waveguide amplifier fabricated by Ag-Na ion-exchange. *Electron. Lett.* **2006**, *42*, 632–633. [CrossRef]

37. Jaouën, Y.; du Mouza, L.; Barbier, D.; Delavaux, J.-M.; Bruno, P. Eight-Wavelength Er–Yb Doped Amplifier: Combiner/Splitter Planar Integrated Module. *IEEE Photonics Technol. Lett.* **1999**, *11*, 1105–1107. [CrossRef]

38. Conzone, S.; Hayden, J.S.; Funk, D.S.; Roshko, A.; Veasey, D.L. Hybrid glass substrates for waveguide device manufacture. *Opt. Lett.* **2001**, *26*, 509–511. [CrossRef]

39. Yliniemi, S.; Albert, J.; Wang, Q.; Honkanen, S. UV-exposed Bragg gratings for laser applications in silver-sodium phosphate glass waveguides. *Opt. Express* **2006**, *14*, 2898–2903. [CrossRef] [PubMed]

40. Molina, G.; Murica, S.; Molera, J.; Roldan, C.; Crespo, D.; Pradell, T. Color and dichroism of silver-stained glasses. *J. Nanopart. Res.* **2013**, *15*, 1932–1937. [CrossRef]

41. Puche-Roig, A.; Martín, V.P.; Murcia- Mascarós, S.; Ibáñez Puchades, R. Float glass colouring by ion-exchange. *J. Cult. Herit.* **2008**, *9*, e129–e133. [CrossRef]

42. Berg, K.-J.; Berger, K.A.; Hofmeister, H. Small silver particles in glass surface layers produced by sodium-silver ion-exchange—their concentration and size depth profile. *Z. Phys. D At. Mol. Clust.* **1991**, *20*, 309–311. [CrossRef]

43. Berger, A. Concentration and size depth profile of colloidal silver particles in glass surface produced by sodium-silver ion-exchange. *J. Non Cryst. Solids* **1992**, *151*, 88–94. [CrossRef]

44. Manikandan, D.; Mohan, S.; Magudapathy, P.; Nair, K.G.M. Blue shift of plasmon resonance in Cu and Ag ion-exchanged and annealed soda-lime glass: An optical absorption study. *Phys. B Condens. Matter.* **2003**, *325*, 86–91. [CrossRef]

45. Quaranta, A.; Rahman, A.; Mariotto, G.; Maurizio, C.; Trave, E.; Gonella, F.; Cattaruzza, E.; Gibaudo, E.; Broquin, J.E. Spectroscopic Investigation of Structural Rearrangements in Silver Ion-Exchanged Silicate Glasses. *J. Phys. Chem. C* **2012**, *116*, 3757–3764. [CrossRef]

46. Mohapatra, S. Tunable surface plasmon resonance of silver nanoclusters in ion-exchanged soda lime glass. *J. Alloys Compd.* **2014**, *598*, 11–15. [CrossRef]

47. Madrigal, J.B.; Tellez-Limon, R.; Gardillou, F.; Barbier, D.; Geng, W.; Couteau, C.; Salas-Montiel, R.; Blaize, S. Hybrid integrated optical waveguides in glass for enhanced visible photoluminescence of nanoemitters. *Appl. Opt.* **2016**, *55*, 10263–10268. [CrossRef] [PubMed]

48. Ashley, P.R.; Bloemer, M.J.; Davis, J.H. Measurement of nonlinear properties in Ag-ion-exchange waveguides using degenerate four-wave mixing. *Appl. Phys. Lett.* **1990**, *57*, 1488–1490. [CrossRef]

49. Faccio, D.; Trapani, D.P.; Borsella, E.; Gonella, F.; Mazzoldi, P.; Malvezzi, A.M. Measurement of the third-order nonlinear susceptibility of Ag nanoparticles in glass in a wide spectral range. *Europhys. Lett.* **1998**, *43*, 213–218. [CrossRef]

50. Tagantsev, D.K.; Kazansky, P.G.; Lipovskii, A.A.; Maluev, K.D. Ion-exchange-induced formation of glassy electrooptical and nonlinear optical nanomaterial. *J. Non Cryst. Solids* **2008**, *354*, 1369–1372. [CrossRef]

51. Karvonen, L.; Rönn, J.; Kujala, S.; Chen, Y.; Säynätjoki, A.; Tervonen, A.; Svirko, Y.; Honkanen, S. High non-resonant third-order optical nonlinearity of Ag–glass nanocomposite fabricated by two-step ion-exchange. *Opt. Mater.* **2013**, *36*, 328–332. [CrossRef]

52. Xiang, W.; Gao, H.; Ma, L.; Ma, X.; Huang, Y.; Pei, L.; Liang, X. Valence State Control and Third-Order Nonlinear Optical Properties of Copper Embedded in Sodium Borosilicate Glass. *ACS Appl. Mater. Interfaces* **2015**, *7*, 10162–10168. [CrossRef] [PubMed]

53. Kumar, P.; Mathpal, M.C.; Hamad, S.; Rao, S.V.; Neethling, J.H.; Janse, A.; van Vuuren, A.J.; Njoroge, E.G.; Kroon, R.E.; Roos, W.D.; et al. Cu nanoclusters in ion-exchanged soda-lime glass: Study of SPR and nonlinear optical behavior for photonics. *Appl. Mater. Today* **2019**, *15*, 323–334. [CrossRef]

54. Malta, O.; Santa-Cruz, P.A.; de Sa´, G.F.; Auzel, F. Fluorescence enhancement induced by the presence of small silver particles in Eu^{3+} doped materials. *J. Lumin.* **1985**, *33*, 261–272. [CrossRef]

55. Hayakawa, T.; Selvan, S.T.; Nogami, M. Field enhancement effect of small Ag particles on the fluorescence from Eu^{3+}-doped SiO$_2$ glass. *Appl. Phys. Lett.* **1999**, *74*, 1513–1515. [CrossRef]

56. Strohhöfer, C.; Polman, A. Silver as a sensitizer for erbium. *Appl. Phys. Lett.* **2002**, *81*, 1414–1416. [CrossRef]

57. Chiasera, A.; Ferrari, M.; Mattarelli, M.; Montagna, M.; Pelli, S.; Portales, H.; Zheng, J.; Righini, G.C. Assessment of spectroscopic properties of erbium ions in a soda-lime silicate glass after silver–sodium exchange. *Opt. Mater.* **2005**, *27*, 1743–1747. [CrossRef]

58. Mattarelli, M.; Montagna, M.; Vishnubhatla, K.; Chiasera, A.; Ferrari, M.; Righini, G.C. Mechanisms of Ag to Er energy transfer in silicate glasses: A photoluminescence study. *Phys. Rev. B* **2007**, *75*, 125102. [CrossRef]

59. Li, L.; Yang, Y.; Zhou, D.; Yang, Z.; Xu, X.; Qiu, J. Investigation of the role of silver species on spectroscopic features of Sm^{3+}-activated sodium aluminosilicate glasses via Ag$^+$-Na$^+$ ion-exchange. *J. Appl. Phys.* **2013**, *113*, 193103. [CrossRef]

60. Li, L.; Yang, Y.; Zhou, D.; Yang, Z.; Xu, X.; Qiu, J. Investigation of the interaction between different types of Ag species and europium ions in Ag$^+$-Na$^+$ ion-exchange glass. *Opt. Mater. Express* **2013**, *3*, 806–812. [CrossRef]

61. Gonella, F. Silver doping glasses. *Ceram. Int.* **2015**, *41*, 6693–6701. [CrossRef]

62. Stanek, S.; Nekvindova, P.; Svecova, B.; Vytykacova, S.; Mika, M.; Oswald, J.; Barkman, O.; Spirkova, J. The influence of silver ion-exchange on the luminescence properties of Er-Yb silicate glasses. *Opt. Mater.* **2017**, *72*, 183–189. [CrossRef]

63. Zhao, J.; Yang, Z.; Yu, C.; Qiu, J.; Song, Z. Preparation of ultra-small molecule-like Ag nano-clusters in silicate glass based on ion-exchange process: Energy transfer investigation from molecule-like Ag nano-clusters to Eu^{3+} ions. *Chem. Eng. J.* **2018**, *341*, 175–186. [CrossRef]

64. Zhao, J.; Zhu, J.; Yang, Z.; Jiao, Q.; Yu, C.; Qiu, J.; Song, Z. Selective preparation of Ag species on photoluminescence of Sm^{3+} in borosilicate glass via Ag^+-Na^+ ion-exchange. *J. Am. Ceram. Soc.* **2019**, *103*, 955–964. [CrossRef]
65. Mironov, L.Y.; Sgibnev, Y.M.; Kolesnikov, I.E.; Nikonorov, N.V. Luminescence and energy transfer mechanisms in photo-thermo-refractive glasses co-doped with silver molecular clusters and Eu^{3+}. *Phys. Chem. Chem. Phys.* **2020**, *22*, 23342–23350. [CrossRef]
66. Sgibnev, Y.; Asamoah, B.; Nikonorov, N.; Honkanen, S. Tunable photoluminescence of silver molecular clusters formed in Na^+-Ag^+ ion-exchanged antimony-doped photo-thermo-refractive glass matrix. *J. Lumin.* **2020**, *226*, 117411. [CrossRef]
67. Allsopp, B.L.; Orman, R.; Johnson, S.R.; Baistow, I.; Sanderson, G.; Sundberg, P.; Stålhandske, C.; Grund, L.; Andersson, A.; Booth, J.; et al. Towards improved cover glasses for photovoltaic devices. *Prog. Photovolt. Res. Appl.* **2020**, *28*, 1187–1206. [CrossRef]
68. Huang, X.; Han, S.; Huang, W.; Liu, X. Enhancing solar cell efficiency: The search for luminescent materials as spectral converters. *Chem. Soc. Rev.* **2013**, *42*, 173–201. [CrossRef] [PubMed]
69. Cattaruzza, E.; Mardegan, M.; Pregnolato, T.; Ungaretti, G.; Aquilanti, G.; Quaranta, A.; Battaglin, G.; Trave, E. Ion-exchange doping of solar cell coverglass for sunlight down-shifting. *Sol. Energy Mater. Sol. Cells* **2014**, *130*, 272–280. [CrossRef]
70. Cattaruzza, E.; Caselli, V.M.; Mardegan, M.; Gonella, F.; Bottaro, G.; Quaranta, A.; Valotto, G.; Enrichi, F. $Ag^+$$\leftrightarrow$$Na^+$ ion-exchanged silicate glasses for solar cells covering: Down-shifting properties. *Ceram. Int.* **2015**, *41*, 7221–7226. [CrossRef]
71. Mardegan, M.; Cattaruzza, E. Cu-doped photovoltaic glasses by ion-exchange for sunlight down–shifting. *Opt. Mater.* **2016**, *61*, 105–110. [CrossRef]
72. Li, Y.; Chen, F.; Liu, C.; Han, J.; Zhao, X. UV–Visible spectral conversion of silver ion-exchanged aluminosilicate glasses. *J. Non-Cryst. Solids* **2017**, *471*, 82–90. [CrossRef]
73. Enrichi, F.; Armellini, C.; Battaglin, G.; Belluomo, F.; Belmokhtar, S.; Bouajaj, A.; Cattaruzza, E.; Ferrari, M.; Gonella, F.; Lukowiak, A.; et al. Silver doping of silica-hafnia waveguides containing Tb^{3+}/Yb^{3+} rare earths for down conversion in PV solar cells. *Opt. Mater.* **2016**, *60*, 264–269. [CrossRef]
74. Sgibnev, Y.; Cattaruzza, E.; Dubrovin, V.; Vasilyev, V.; Nikonorov, N. Photo-Thermo-Refractive Glasses Doped with Silver Molecular Clusters as Luminescence Downshifting Material for Photovoltaic Applications. *Part. Part. Syst. Charact.* **2018**, *35*, 1800141. [CrossRef]
75. Bubli, I.; Ali, S.; Ali, M.; Hayat, K.; Iqbal, Y.; Zulfiqar, S.; ul Haq, A.; Cattaruzza, E. Enhancement of solar cell efficiency via luminescent downshifting by an optimized coverglass. *Ceram. Int.* **2020**, *46*, 2110–2115. [CrossRef]
76. Enrichi, F.; Belmokhtar, S.; Benedetti, A.; Bouajaj, A.; Cattaruzza, E.; Coccetti, F.; Colusso, E.; Ferrari, M.; Ghamgosar, P.; Gonella, F.; et al. Ag nanoaggregates as efficient broadband sensitizers for Tb^{3+} ions in silica-zirconia ion-exchanged sol-gel glasses and glass-ceramics. *Opt. Mater.* **2018**, *84*, 668–674. [CrossRef]
77. Tresnakova, P.; Malichova, H.; Spirkova, J.; Mika, M. Fabrication of optical layers containing Er(III) and Cu(I) in silicate glasses. *J. Phys. Chem. Solids* **2007**, *68*, 1276–1279. [CrossRef]
78. Yu, C.; Zhao, J.; Zhu, J.; Huang, A.; Qiu, J.; Song, Z.; Zhou, D. Luminescence enhancement and white light generation of Eu^{3+} and Dy^{3+} single-doped and co-doped tellurite glasses by Ag nanoparticles based on Ag^+-Na^+ ion-exchange. *J. Alloys Compd.* **2018**, *748*, 717–729. [CrossRef]
79. Tang, L.; Chen, N. White light emitting YVO_4:Eu^{3+},Tm^{3+},Dy^{3+} nanometer- and submicrometer-sized particles prepared by an ion-exchange method. *Ceram. Int.* **2016**, *42*, 302–309. [CrossRef]
80. Sgibnev, Y.M.; Nikonorov, N.V.; Ignatiev, A.I. Luminescence of silver clusters in ion-exchanged cerium-doped photo-thermo-refractive glasses. *J. Lumin.* **2016**, *176*, 292–297. [CrossRef]
81. Sgibnev, Y.M.; Nikonorov, N.V.; Ignatiev, A.I. High efficient luminescence of silver clusters in ion-exchanged antimony doped photo-thermo-refractive glasses: Influence of antimony content and heat treatment parameters. *J. Lumin.* **2017**, *188*, 172–179. [CrossRef]
82. Zhou, Y.; Laybourn, P.J.R.; Magill, J.V.; De La Rue, R.M. An evanescent fluorescence biosensor using ion-exchanged buried waveguides and the enhancement of peak fluorescence. *Biosens. Bioelectron.* **1991**, *6*, 595–607. [CrossRef]
83. Zhou, Y.; Magill, J.V.; De La Rue, R.M.; Laybourn, P.J.R. Evanescent fluorescence immunoassays performed with a disposable ion-exchanged patterned waveguide. *Sens. Actuators B Chem.* **1993**, *11*, 245–250. [CrossRef]
84. Srivastava, R.; Bao, C.; Gómez-Reino, C. Planar-surface waveguide evanescent-wave chemical sensors. *Sens. Actuators A Phys.* **1996**, *51*, 165–171. [CrossRef]
85. Hassanzadeh, A.; Nitsche, M.; Mittler, S.; Armstrong, S.; Dixon, S.J.; Langbein, U. Waveguide evanescent field fluorescence microscopy: Thin film fluorescence intensities and its application in cell biology. *Appl. Phys. Lett.* **2008**, *92*, 233503. [CrossRef]
86. Hassanzadeh, A.; Nitsche, M.; Armstrong, S.; Nabavi, N.; Harrison, R.; Dixon, S.J.; Langbein, U. Optical waveguides formed by silver ion-exchange in Schott SG11 glass for waveguide evanescent field fluorescence microscopy: Evanescent images of HEK293 cells. *J. Biomed. Opt.* **2010**, *15*, 036018. [CrossRef]
87. Araci, I.E.; Mendes, S.B.; Yurt, N.; Honkanen, S.; Peyghambarian, N. Highly sensitive spectroscopic detection of heme-protein sub-monolayer films by channel integrated optical waveguide. *Opt. Express* **2007**, *15*, 5595–5603. [CrossRef] [PubMed]
88. Karabchevsky, A.; Kavokin, A.V. Giant absorption of light by molecular vibrations on a chip. *Sci. Rep.* **2016**, *6*, 21201. [CrossRef]
89. Zhu, M.; Kari, N.; Yan, Y.; Yimit, A. The fabrication and gas sensing application of a fast-responding m-CP-PVP composite film/potassium ion-exchanged glass optical waveguide. *Anal. Methods* **2017**, *9*, 5494–5501. [CrossRef]

90. Du, B.; Tong, Z.; Mu, X.; Xu, J.; Liu, S.; Liu, Z.; Cao, W.; Qi, Z.-M. A Potassium Ion-Exchanged Glass Optical Waveguide Sensor Locally Coated with a Crystal Violet-SiO2 Gel Film for Real-Time Detection of Organophosphorus Pesticides Simulant. *Sensors* **2019**, *19*, 4219. [CrossRef] [PubMed]

91. Homola, J.; Čtyroký, J.; Skalský, M.; Hradilová, J.; Kolářová, J.P. A surface plasmon resonance based integrated optical sensor. *Sens. Actuators B Chem.* **1997**, *39*, 286–290. [CrossRef]

92. Dostálek, J.; Čtyroký, J.; Homola, J.; Brynda, E.; Skalský, M.; Nekvindová, P.; Špirková, J.; Škvor, J.; Schröfel, J. Surface plasmon resonance biosensor based on integrated optical waveguide. *Sens. Actuators B Chem.* **2001**, *76*, 8–12. [CrossRef]

93. De Bonnault, S.; Bucci, D.; Zermatten, P.J.; Charette, P.G.; Broquin, J.E. Hybrid metallic ion-exchanged waveguides for SPR biological sensing. In Proceedings of the SPIE, San Francisco, CA, USA, 7–12 February 2015; Volume 9365, p. 93650. [CrossRef]

94. Tellez-Limon, R.; Gardillou, F.; Coello, V.; Salas-Montiel, R. Coupled localized surface plasmon resonances in periodic arrays of gold nanowires on ion-exchange waveguide technology. *J. Opt.* **2021**, *23*, 25801. [CrossRef]

95. Nashchekin, A.V.; Nevedomskiy, V.N.; Obraztsov, P.A.; Stepanenko, O.V.; Sidorov, A.I.; Usov, O.A.; Turoverov, K.K.; Konnikov, S.G. Waveguide-type localized plasmon resonance biosensor for noninvasive glucose concentration detection. In Proceedings of the SPIE, Brussels, Belgium, 16–19 April 2012; Volume 8427, p. 842739. [CrossRef]

96. Pilot, R.; Signorini, R.; Durante, C.; Orian, L.; Bhamidipati, M.; Fabris, L. A Review on Surface-Enhanced Raman Scattering. *Biosensors* **2019**, *9*, 57. [CrossRef] [PubMed]

97. Mosier-Boss, P.A. Review of SERS Substrates for Chemical Sensing. *Nanomaterials* **2017**, *7*, 142. [CrossRef]

98. Zhang, J.; Dong, W.; Sheng, J.; Zheng, J.; Li, J.; Qiao, L.; Jiang, L. Silver nanoclusters formation in ion-exchanged glasses by thermal annealing, UV-laser and X-ray irradiation. *J. Non Cryst. Solids* **2008**, *310*, 234–239. [CrossRef]

99. Kolobkova, E.; Sergeevna Kuznetsova, M.; Nikonorov, N. Ag/Na Ion-exchange in Fluorophosphate Glasses and Formation of Ag Nanoparticles in the Bulk and on the Surface of the Glass. *ACS Appl. Nano Mater.* **2019**, *2*, 6928–6938. [CrossRef]

100. Karvonen, L.; Chen, Y.; Säynätjoki, A.; Taiviola, K.; Tervonen, A.; Honkanen, S. SERS-active silver nanoparticle aggregates produced in high-iron float glass by ion-exchange process. *Opt. Mater.* **2011**, *34*, 1–5. [CrossRef]

101. Babich, E.; Kaasik, V.; Redkov, A.; Maurer, T.; Lipovskii, A. SERS-Active Pattern in Silver-Ion-Exchange Glass Drawn by Infrared Nanosecond Laser. *Nanomaterials* **2020**, *10*, 1849. [CrossRef] [PubMed]

102. Tite, T.; Ollier, N.; Sow, M.C.; Vocanson, F.; Goutaland, F. Ag nanoparticles in soda-lime glass grown by continuous wave laser irradiation as an efficient SERS platform for pesticides detection. *Sens. Actuators B Chem.* **2017**, *242*, 127–131. [CrossRef]

103. Chen, Y.; Jaakola, J.; Ge, Y.; Säynätjoki, A.; Tervonen, A.; Hannula, S.-P.; Honkanen, S. In situ fabrication of waveguide-compatible glass-embedded silver nanoparticle patterns by masked ion-exchange process. *J. Non Cryst. Solids* **2009**, *355*, 2224–2227. [CrossRef]

104. Chen, Y.; Jaakola, J.; Säynätjoki, A.; Tervonen, A.; Honkanen, S. Glass-embedded silver nanoparticle patterns by masked ion-exchange process for surface-enhanced Raman scattering. *J. Raman Spectrosc.* **2011**, *42*, 936–940. [CrossRef]

105. Simo, A.; Joseph, V.; Fenger, R.; Kneipp, J.; Rademann, K. Long-Term Stable Silver Subsurface Ion-Exchanged Glasses for SERS Applications. *Chem. Phys. Chem.* **2011**, *12*, 1683–1688. [CrossRef]

106. Simo, A.; Polte, J.; Pfänder, N.; Vainio, U.; Emmerling, F.; Rademann, K. Formation Mechanism of Silver Nanoparticles Stabilized in Glassy Matrices. *J. Am. Chem. Soc.* **2012**, *134*, 18824–18833. [CrossRef]

107. Hofmeister, H.; Tan, G.L.; Dubiel, M. Shape and internal structure of silver nanoparticles embedded in glass. *J. Mater. Res.* **2005**, *20*, 1551–1562. [CrossRef]

108. Williams, T. *The Glass Menagerie*, 1st ed.; Random House Publisher: New York, NY, USA, 1945.

109. Chesterton, G.K. *Orthodoxy*, 1st ed.; John Lane/Bodley Head Publisher: London, UK, 1909.

110. Varshneya, A.K. Chemical Strengthening of Glass: Lessons Learned and Yet to Be Learned. *Int. J. Appl. Glass Sci.* **2010**, *1*, 131–142. [CrossRef]

111. Romaniuk, R.S. Tensile strength of tailored optical fibres. *Opto Electron. Rev.* **2000**, *8*, 101–116.

112. Wisitsorasak, A.; Wolynes, P. On the strength of glasses. *PNAS* **2012**, *109*, 16068–16072. [CrossRef] [PubMed]

113. Nielsen, J.H.; Bjarrum, M. Deformations and strain energy in fragments of tempered glass: Experimental and numerical investigation. *Glass Struct. Eng.* **2017**, *2*, 133–146. [CrossRef]

114. Datsiou, K.C.; Overend, M. The strength of aged glass. *Glass Struct. Eng.* **2017**, *2*, 105–120. [CrossRef]

115. Gardon, R. Thermal tempering of glass. In *Glass Science and Technology. Elasticity and Strength in Glasses*; Uhlmann, D.R., Kreidl, N.J., Eds.; Academic Press: New York, NY, USA, 1980; Volume 5, pp. 145–216. [CrossRef]

116. Sglavo, V.M.; Quaranta, A.; Allodi, V.; Mariotto, G. Analysis of surface structure of soda lime silicate glasses after chemical strengthening in different KNO3 salt bath. *J. Non Cryst. Solids* **2014**, *401*, 105–109. [CrossRef]

117. Mognato, E.; Schiavonato, M.; Barbieri, A.; Pittoni, M. Process influences on mechanical strength of chemical strengthened glass. *Glass Struct. Eng.* **2016**, *1*, 247–260. [CrossRef]

118. Rogoziński, R. Stresses Produced in the BK7 Glass by the K$^+$–Na$^+$ Ion-exchange: Real-Time Process Control Method. *Appl. Sci.* **2019**, *9*, 2548. [CrossRef]

119. Cooper, A.R.; Krohn, D.A. Strengthening of glass fibres II: Ion-exchange. *J. Am. Ceram. Soc.* **1969**, *52*, 665–669. [CrossRef]

120. Dugnani, R. Residual stress in ion-exchanged silicate glass: An analytical solution. *J. Non Cryst. Solids* **2017**, *471*, 368–378. [CrossRef]

121. Varshneya, A.K. The physics of chemical strengthening of glass: Room for a new view. *J. Non Cryst. Solids* **2010**, *356*, 2289–2294. [CrossRef]

122. Macrelli, G. Chemically strengthened glass by ion-exchange: Strength evaluation. *Int. J. Appl. Glass Sci.* **2017**, *9*, 156–166. [CrossRef]
123. Macrelli, G.; Varshneya, A.K.; Mauro, J.C. Simulation of glass network evolution during chemical strengthening: Resolution of the subsurface compression maximum anomaly. *J. Non Cryst. Solids* **2019**, *552*, 119457. [CrossRef]
124. Macrelli, G.; Özben, N.; Kayaalp, A.C.; Ersundu, M.C.; Sökmen, I. Stress in ion-exchanged soda-lime silicate and sodium aluminosilicate glasses: Experimental and theoretical comparison. *Int. J. Appl. Glass Sci.* **2020**, *11*, 730–742. [CrossRef]
125. Terakado, N.; Sasaki, R.; Takahashi, Y.; Fujiwara, T.; Orihara, S.; Orihara, Y. A novel method for stress evaluation in chemically strengthened glass based on micro-Raman spectroscopy. *Commun. Phys.* **2020**, *3*, 1–7. [CrossRef]
126. Erdem, I.; Guldiren, D.; Aydin, S. Chemical tempering of soda lime silicate glasses by ion-exchange process for the improvement of surface and bulk mechanical strength. *J. Non Cryst. Solids* **2017**, *473*, 170–178. [CrossRef]
127. Aaldenberg, E.M.; Lezzi, P.J.; Seaman, J.H.; Blanchet, T.A.; Tomozawa, M. Ion-Exchanged Lithium Aluminosilicate Glass: Strength and Dynamic Fatigue. *J. Am. Ceram. Soc.* **2016**, *99*, 2645–2654. [CrossRef]
128. Li, X.; Jiang, L.; Wang, Y.; Mohagheghian, I.; Dear, J.P.; Ki, L.; Yan, Y. Correlation between K^+-Na^+ diffusion coefficient and flexural strength of chemically tempered aluminosilicate glass. *J. Non Cryst. Solids* **2017**, *47*, 72–81. [CrossRef]
129. Ragoen, C.; Sen, S.; Lambricht, T.; Godet, S. Effect of Al2O3 content on the mechanical and interdiffusional properties of ion-exchanged Na-aluminosilicate glasses. *J. Non Cryst. Solids* **2017**, *458*, 129–136. [CrossRef]
130. Available online: https://www.corning.com/microsites/csm/gorillaglass/PI_Sheets/2020/Corning%20Gorilla%20Glass%20 6_PI%20Sheet.pdf (accessed on 17 May 2021).
131. Available online: https://photonicsolutioncenter.com/media/companies/133/products/439/documents/agc-dragontrail-chemically.pdf (accessed on 17 May 2021).
132. Available online: https://www.schott.com/d/xensation/7d59e8ca-a167-4d23-b375-6a8040cc9eb6/1.0/schott-xensation-up-datasheet-english-row-20052019.pdf (accessed on 17 May 2021).
133. Available online: https://abrisatechnologies.com/specs/SCHOTT%20Xensation%20Spec%20Sheet.pdf#zoom=100 (accessed on 17 May 2021).
134. Wilson, J.R.; Jones, C.D.; Bartlow, C.C.; Doan, A.R. Patterned Asymmetric Chemical Strengthening. US Patent 20200017406A1, 16 January 2020.
135. Luzzato, V.; Prest, C.D.; Memering, D.N.; Rogers, M.S. Asymmetric Chemical Strengthening. US Patent 20170334769A1, 23 November 2017.
136. Rogers, M.S.; Memering, D.N.; Prest, C.D. Asymmetric Chemical Strengthening. US Patent 20150274585A1, 1 October 2015.
137. Available online: https://www.corning.com/microsites/csm/gorillaglass/PI_Sheets/2020/Corning%20Gorilla%20Glass%20 Victus_PI%20Sheet.pdf (accessed on 17 May 2021).
138. Available online: https://www.apple.com/iphone-12/ (accessed on 17 May 2021).
139. Available online: https://media.ford.com/content/fordmedia/fna/us/en/news/2015/12/15/industry-first-gorilla-glass-hybrid-windshield-on-all-new-ford-GT.html (accessed on 17 May 2021).
140. Available online: https://phys.org/news/2015-12-gorilla-glass-cell-cars.html (accessed on 17 May 2021).
141. Wang, H.F.; Xing, G.Z.; Wang, X.Y.; Zhang, L.L.; Zhang, L.; Li, S. Chemically strengthened protection glasses for the applications of space solar cells. *AIP Adv.* **2014**, *4*, 47133. [CrossRef]
142. Kambe, M.; Hara, K.; Mitarai, K.; Takeda, S.; Fukawa, M.; Ishimaru, N.; Kondo, M. Chemically strengthened cover glass for preventing potential induced degradation of crystalline silicon solar cells. In Proceedings of the 39th Photovoltaic Specialists Conference, Tampa Bay, FL, USA, 16–21 June 2013. [CrossRef]
143. Li, J.; Wei, R.; Liu, X.; Guo, H. Enhanced luminescence via energy transfer from Ag^+ to RE ions (Dy^{3+}, Sm^{3+}, Tb^{3+}) in glasses. *Opt. Express* **2012**, *20*, 10122–10127. [CrossRef]
144. Jiao, Q.; Wang, X.; Qiu, J.; Zhou, D. Effect of silver ions and clusters on the luminescence properties of Eu-doped borate glasses. *Mater. Res. Bull.* **2015**, *72*, 264–268. [CrossRef]
145. Li, L.; Yang, Y.; Zhou, D.; Xu, X.; Qiu, J. The influence of Ag species on spectroscopic features of Tb^{3+}-activated sodium–aluminosilicate glasses via Ag^+–Na^+ ion exchange. *J. Non Cryst. Solids* **2014**, *385*, 95–99. [CrossRef]
146. Bozelli, J.C.; de Oliveira Nunes, L.A.; Sigoli, F.A.; Mazali, I.O. Erbium and Ytterbium Codoped Titanoniobophosphate Glasses for Ion-Exchange-Based Planar Waveguides. *J. Am. Ceram. Soc.* **2010**, *93*, 2689–2692. [CrossRef]
147. Vařák, P.; Vytykáčová, S.; Nekvindová, P.; Michalcová, A.; Malinský, P. The influence of copper and silver in various oxidation states on the photoluminescence of Ho^{3+}/Yb^{3+} doped zinc-silicate glasses. *Opt. Mater.* **2019**, *91*, 253–260. [CrossRef]
148. He, X.; Xu, X.; Shi, Y.; Qiu, J. Effective enhancement of Bi near-infrared luminescence in silicogermanate glasses via silver–sodium ion exchange. *J. Non Cryst. Solids* **2015**, *409*, 178–182. [CrossRef]
149. Paje, S.E.; García, M.A.; Villegas, M.A.; Llopis, J. Cerium doped soda-lime-silicate glasses: Effects of silver ion-exchange on optical properties. *Opt. Mater.* **2001**, *17*, 459–469. [CrossRef]
150. Grand, J.; Lamy de la Chapelle, M.; Bijeon, J.-L.; Adam, P.-M.; Vial, A.; Royer, P. Role of localized surface plasmons in surface-enhanced Raman scattering of shape-controlled metallic particles in regular arrays. *Phys. Rev. B* **2005**, *72*, 033407. [CrossRef]
151. Li, P.; Long, F.; Chen, W.; Chen, J.; Chu, P.K.; Wang, H. Fundamentals and applications of surface-enhanced Raman spectroscopy-based biosensors. *Curr. Opin. Biomed. Eng.* **2020**, *13*, 51–59. [CrossRef]

152. Betz, J.F.; Yu, W.W.; Cheng, Y.; White, I.M.; Rubloff, G.W. Simple SERS substrates: Powerful, portable, and full of potential. *PCCP* **2014**, *16*, 2224–2239. [CrossRef]

153. Fan, M.; Andrade, G.F.S.; Brolo, A.G. A review on the fabrication of substrates for surface enhanced Raman spectroscopy and their applications in analytical chemistry. *Anal. Chim. Acta* **2011**, *693*, 7–25. [CrossRef] [PubMed]

154. Zhao, J.; Lin, J.; Zhang, W.; Wei, H.; Chen, Y. SERS-active Ag nanoparticles embedded in glass prepared by a two-step electric field-assisted diffusion. *Opt. Mater.* **2014**, *39*, 97–102. [CrossRef]

155. Chen, Y.; Jaakola, J.; Säynätjoki, A.; Tervonen, A.; Honkanen, S. SERS-active silver nanoparticles in ion-exchanged glass. *J. Nonlinear Opt. Phys.* **2010**, *19*, 527–533. [CrossRef]

156. Manikandan, P.; Manikandan, D.; Manikandan, E.; Christy Ferdinand, A. Surface enhanced Raman scattering (SERS) of silver ions embedded nanocomposite glass. *Spectrochim. Acta A Mol. Biomol. Spectrosc.* **2014**, *124*, 203–207. [CrossRef]

157. Manikandan, P.; Manikandan, D.; Manikandan, E.; Christy Ferdinand, A. Structural, Optical and Micro-Raman Scattering Studies of Nanosized Copper Ion (Cu+) Exchanged Soda Lime Glasses. *Plasmonics* **2014**, *9*, 637–643. [CrossRef]

158. Quaranta, A.; Cattaruzza, E.; Gonella, F.; Rahman, A.; Mariotto, G. Cross-sectional Raman micro-spectroscopy study of silver nanoparticles in soda–lime glasses. *J. Non Cryst. Solids* **2014**, *401*, 219–223. [CrossRef]

159. Chen, Y.; Karvonen, L.; Säynätjoki, A.; Ye, C.; Tervonen, A.; Honkanen, S. Ag nanoparticles embedded in glass by two-step ion exchange and their SERS application. *Opt. Mater. Express* **2011**, *1*, 164–172. [CrossRef]

160. Manzani, D.; Franco, D.F.; Afonso, C.R.M.; Sant'Ana, A.C.; Nalih, M.; Ribeiro, S.J.L. A new SERS substrate based on niobium lead-pyrophosphate glasses obtained by Ag+/Na+ ion exchange. *Sens. Actuators B Chem.* **2018**, *277*, 347–352. [CrossRef]

161. De Marchi, G.; Caccavale, F.; Gonella, F.; Mattei, G.; Mazzoldi, P.; Battaglin, G.; Quaranta, A. Silver nanoclusters formation in ion-exchanged waveguides by annealing in hydrogen atmosphere. *Appl. Phys. A* **1996**, *63*, 403–407. [CrossRef]

162. Redkov, A.V.; Zhurikhina, V.V.; Lipovskii, A.A. Formation and self-arrangement of silver nanoparticles in glass via annealing in hydrogen: The model. *J. Non Cryst. Solids* **2013**, *376*, 152–157. [CrossRef]

163. Chervinskii, S.; Sevriuk, V.; Reduto, I.; Lipovskii, A. Formation and 2D-patterning of silver nanoisland film using thermal poling and out-diffusion from glass. *J. Appl. Phys.* **2013**, *114*, 224301. [CrossRef]

164. Redkov, A.; Chervinskii, S.; Baklanov, A.; Reduto, I.; Zhurikhina, V.; Lipovskii, A. Plasmonic molecules via glass annealing in hydrogen. *Nanoscale Res. Lett.* **2014**, *9*, 606. [CrossRef] [PubMed]

165. Goutaland, F.; Sow, M.; Ollier, N.; Vacanson, F. Growth of highly concentrated silver nanoparticles and nanoholes in silver-exchanged glass by ultraviolet continuous wave laser exposure. *Opt. Mater. Express* **2012**, *2*, 350–357. [CrossRef]

166. Goutaland, F.; Colombier, J.-P.; Cherif Sow, M.; Ollier, N.; Vocanson, F. Laser-induced periodic alignment of Ag nanoparticles in soda-lime glass. *Opt. Express* **2013**, *21*, 31789–31799. [CrossRef]

167. Niry, M.D.; Mostafavi-Amjad, J.; Khalesifard, H.R.; Ahangary, A.; Azizian-Kalandaragh, Y. Formation of silver nanoparticles inside a soda-lime glass matrix in the presence of a high intensity Ar+ laser beam. *J. Appl. Phys.* **2012**, *111*, 033111. [CrossRef]

168. Miotello, A.; Bonelli, M.; De Marchi, G.; Mattei, G.; Mazzoldi, P.; Sada, C.; Gonella, F. Formation of silver nanoclusters by excimer–laser interaction in silver-exchanged soda-lime glass. *Appl. Phys. Lett.* **2001**, *79*, 2456–2458. [CrossRef]

169. Sheng, J.; Zheng, J.; Zhang, J.; Zhou, C.; Jiang, L. UV-laser-induced nanoclusters in silver ion-exchanged soda-lime silicate glass. *Physica B* **2007**, *387*, 32–35. [CrossRef]

170. Wackerow, S.; Abdolvand, A. Laser-assisted one-step fabrication of homogeneous glass–silver composite. *Appl. Phys. A* **2012**, *109*, 45–49. [CrossRef]

171. Wackerow, S.; Abdolvand, A. Generation of silver nanoparticles with controlled size and spatial distribution by pulsed laser irradiation of silver ion-doped glass. *Opt. Express* **2014**, *22*, 5076–5085. [CrossRef]

172. Schmidl, G.; Dellith, J.; Schneidewind, H.; Zopf, D.; Stranik, O.; Gawlik, A.; Anders, S.; Tympel, V.; Katzer, C.; Schmidl, F.; et al. Formation and characterization of silver nanoparticles embedded in optical transparent materials for plasmonic sensor surfaces. *Mater. Sci. Eng. B* **2015**, *193*, 207–216. [CrossRef]

173. Chervinskii, S.; Matikainen, A.; Dergachev, A.; Lipovskii, A.A.; Honkanen, S. Out-diffused silver island films for surface-enhanced Raman scattering protected with TiO$_2$ films using atomic layer deposition. *Nanoscale Res. Lett.* **2014**, *9*, 398. [CrossRef]

174. Inácio, P.L.; Barreto, B.J.; Horowitz, F.; Correia, R.R.B.; Pereira, M.B. Silver migration at the surface of ion-exchange waveguides: A plasmonic template. *Opt. Mater. Express* **2013**, *3*, 390–399. [CrossRef]

175. Hu, J.; Li, L.; Lin, H.; Zhang, P.; Zhou, W.; Ma, Z. Flexible integrated photonics: Where materials, mechanics and optics meet. *Opt. Mater. Express* **2013**, *3*, 1313–1331. [CrossRef]

176. Geiger, S.; Michon, J.; Liu, S.; Qin, J.; Ni, J.; Hu, J.; Gu, T.; Lu, N. Flexible and stretchable Photonics: The Next Stretch of Opportunities. *ACS Photonics* **2020**, *7*, 2618–2635. [CrossRef]

177. Righini, G.C.; Szczurek, A.; Krzak, J.; Lukowiak, A.; Ferrari, M.; Varas, S.; Chiasera, A. Flexible Photonics: Where Are We Now? In Proceedings of the 22nd International Conference on Transparent Optical Networks (ICTON), Bari, Italy, 19–23 July 2020; pp. 1–4. [CrossRef]

178. Righini, G.C.; Krzak, J.; Lukowiak, A.; Macrelli, G.; Varas, S.; Ferrari, M. From flexible electronics to flexible photonics: A brief overview. *Opt. Mater.* **2021**, *115*, 111011. [CrossRef]

179. Sayginer, O.; Iacob, E.; Varas, S.; Szczurek, A.; Ferrari, M.; Lukowiak, A.; Righini, G.C.; Bursi, O.S.; Chiasera, A. Design, fabrication and assessment of an optomechanical sensor for pressure and vibration detection using flexible glass multilayers. *Opt. Mater.* **2021**, *115*, 111023. [CrossRef]

180. Chen, Y.; Li, H.; Li, M. Flexible and tunable silicon photonic circuits on plastic substrates. *Sci. Rep.* **2012**, *2*, 622. [CrossRef]
181. Fan, L.; Varghese, L.T.; Xuan, Y.; Wang, J.; Niu, B.; Qi, M. Direct fabrication of silicon photonic devices on a flexible platform and its application for strain sensing. *Opt. Express* **2012**, *20*, 20564–20575. [CrossRef]
182. Chiasera, A.; Sayginer, O.; Iacob, E.; Szczurek, A.; Varas, S.; Krzak, J.; Bursi, O.S.; Zonta, D.; Lukowiak, A.; Righini, G.C.; et al. Flexible photonics: RF-sputtering fabrication of glass-based systems operating under mechanical deformation conditions. In Proceedings of the SPIE, Strasbourg, France, 29 March–2 April 2020; Volume 11357, p. 1135705. [CrossRef]
183. Li, L.; Lin, H.; Qiao, S.; Zou, Y.; Danto, S.; Richardson, K.; Musgraves, J.D.; Lu, N.; Hu, J. Integrated flexible chalcogenide glass photonic devices. *Nature Photon.* **2014**, *8*, 643–649. [CrossRef]
184. Li, L.; Lin, H.; Qiao, S.; Huang, Y.-H.; Li, J.-Y.; Michon, J.; Gu, T.; Alosno-Ramos, C.; Vivien, L.; Yadav, A.; et al. Monolithically integrated stretchable photonics. *Light Sci. Appl.* **2018**, *7*, 17138. [CrossRef]
185. Li, L.; Lin, H.; Michon, J.; Huang, Y.; Li, J.; Du, Q.; Yadav, A.; Richardson, K.; Gu, T.; Hu, J. A new twist on glass: A brittle material enabling flexible integrated photonics. *Int. J. Appl. Glass Sci.* **2017**, *8*, 61–68. [CrossRef]
186. Lapointe, J.; Ledemi, Y.; Loranger, S.; Iezzi, V.L.; De Lima Filho, E.S.; Parent, F.; Morency, S.; Messaddeq, Y.; Kashyap, R. Fabrication of ultrafast laser written low-loss waveguides in flexible As$_2$S$_3$ chalcogenide glass tape. *Opt. Lett.* **2016**, *41*, 203–206. [CrossRef] [PubMed]
187. Available online: https://www.corning.com/worldwide/en/innovation/corning-emerging-innovations/corning-willow-glass.html (accessed on 17 May 2021).
188. Available online: https://www.schott.com/en-gb/products/af-32-eco (accessed on 17 May 2021).
189. Hou, Y.; Cheng, J.; Kang, J.; Cui, J. The forming simulation of flexible glass with silt down draw method. *IOP Conf. Ser.: Mater. Sci. Eng.* **2018**, *324*, 12020. [CrossRef]
190. Garner, S.; Glaesemann, S.; Li, X. Ultra-slim flexible glass for roll-to-roll electronic device fabrication. *Appl. Phys. A* **2014**, *116*, 403–407. [CrossRef]
191. Huang, S.; Li, M.; Garner, S.M.; Li, M.-J.; Chen, K.P. Flexible photonic components in glass substrates. *Opt. Express* **2015**, *23*, 22532–22543. [CrossRef] [PubMed]
192. Ellison, A.; Cornejo, I.A. Glass Substrates for Liquid Crystal Displays. *Int. J. Appl. Glass Sci.* **2010**, *1*, 87–103. [CrossRef]
193. Mauro, J.C.; Truesdale, C.M.; Adib, K.; Whittier, A.M.; Martin, A.W. Chemical Strengthening of Alkali-Free Glass via Pressure Vessel Ion Exchange. *Int. J. Appl. Glass Sci.* **2016**, *7*, 446–451. [CrossRef]
194. Martin, A.W.; Mauro, J.C.; Truesdale, C.M. Strengthened Glass, Glass-Ceramic, and Ceramic Articles and Methods of Making the Same through Pressurized Ion Exchange. U.S. Patent Application 2016/0145152 A1, 2 June 2016.
195. Macrelli, G.; Varshneya, A.K.; Mauro, J.C. Ultra-thin glass as a substrate for flexible photonics. *Opt. Mater.* **2020**, *106*, 109994. [CrossRef]
196. Macrelli, G. Strengthened glass by ion exchange, mechanical and optical properties: Perspectives and limits of glass as a substrate for flexible photonics. In Proceedings of the SPIE Photonics Europe, Strasbourg, France, 29 March–2 April 2020; Volume 11357, p. 1135707. [CrossRef]
197. Chiasera, A.; Dumeige, Y.; Féron, P.; Ferrari, M.; Jestin, Y.; Nunzi Conti, G.; Pelli, S.; Soria, S.; Righini, G.C. Spherical whispering-gallery-mode microresonators. *Laser Photonics Rev.* **2010**, *4*, 457–482. [CrossRef]
198. Ward, J.; Benson, O. WGM microresonators: Sensing, lasing and fundamental optics with microspheres. *Laser Photonics Rev.* **2011**, *5*, 553–570. [CrossRef]
199. Strekalov, D.V.; Marquardt, C.; Matsko, A.B.; Schwefel, H.G.L.; Leuchs, G. Nonlinear and quantum optics with whispering gallery resonators. *J. Opt.* **2016**, *18*, 18–123002. [CrossRef]
200. Gorodetsky, M.L.; Savchenkov, A.A.; Ilchenko, V.S. Ultimate Q of optical microsphere resonators. *Opt. Lett.* **1996**, *21*, 453–455. [CrossRef]
201. Ristić, D.; Berneschi, S.; Camerini, M.; Farnesi, D.; Pelli, S.; Trono, C.; Chiappini, A.; Chiasera, A.; Ferrari, M.; Lukowiak, A.; et al. Photoluminescence and lasing in whispering gallery mode glass microspherical resonators. *J. Lumin.* **2016**, *170*, 755–760. [CrossRef]
202. Ward, J.M.; Yang, Y.; Nic Chormaic, S. Glass-on-glass fabrication of bottle-shaped tunable microlasers and their applications. *Sci. Rep.* **2016**, *6*, 25152. [CrossRef] [PubMed]
203. Yu, J.; Lewis, E.; Farrell, G.; Wang, P. Compound glass microsphere resonator devices. *Micromachines* **2018**, *9*, 356. [CrossRef] [PubMed]
204. Zhu, J.; Zohrabi, M.; Bae, K.; Horning, T.M.; Grayson, M.B.; Park, W.; Gopinath, J.T. Nonlinear characterization of silica and chalcogenide microresonators. *Optica* **2019**, *6*, 716–722. [CrossRef]
205. Yu, J.; Zhang, J.; Wang, R.; LI, A.; Zhang, M.; Wang, S.; Wang, P.; Ward, J.M.; Nic Chormaic, S. A tellurite glass optical microbubble resonator. *Opt. Express* **2020**, *28*, 32858–32868. [CrossRef] [PubMed]
206. Fujiwara, H.; Sasaki, K. Up-conversion lasing of a thulium-ion-doped fluorozirconate glass microsphere. *J. Appl. Phys.* **1999**, *86*, 2385–2388. [CrossRef]
207. Von Klitzing, W.; Jahier, E.; Long, R.; Lissillour, F.; Lefèvre-Seguin, V.; Hare, J.; Raimond, J.-M.; Haroche, S. Very low threshold green lasing in microspheres by up-conversion of IR photons. *J. Opt. B Quantum Semiclass. Opt.* **2000**, *2*, 204–206. [CrossRef]
208. Wu, Y.; Ward, J.M.; Nic Chormaic, S. Ultralow threshold green lasing and optical bistability in ZBNA (ZrF$_4$–BaF$_2$–NaF–AlF$_3$) microspheres. *J. Appl. Phys.* **2010**, 33103. [CrossRef]

209. Wang, X.; Yu, Y.; Wang, S.; Ward, J.M.; Nic Chormaic, S.; Wang, P. Single mode green lasing and multicolour luminescent emission from an Er^{3+}-Yb^{3+} co-doped compound fluorosilicate glass microsphere resonator. *OSA Contin.* **2018**, *1*, 261–273. [CrossRef]
210. Huang, Y.; Zhuang, S.; Peng, L.; Wu, J.; Liao, T.; Xu, C.; Duan, Y. Efficiency improvement of up-conversion luminescence of Yb^{3+}/Tm^{3+} co-doped tellurite glass microsphere. *J. Alloys Compd.* **2018**, *748*, 93–99. [CrossRef]
211. Singh, S.K.; Giri, N.K.; Rai, D.K.; Rai, S.B. Enhanced up-conversion emission in Er^{3+}-doped tellurite glass containing silver nanoparticles. *Solid State Sci.* **2010**, *12*, 1480–1483. [CrossRef]
212. Reza Dousti, M. Efficient infrared-to-visible up-conversion emission in Nd^{3+}-doped PbO-TeO_2 glass containing silver nanoparticles. *J. Appl. Phys.* **2013**, *114*, 113105. [CrossRef]
213. Soltani, I.; Hraiech, S.; Horchani-Naifer, K.; Férid, M. Effects of silver nanoparticles on the enhancement of up-conversion and infrared emission in Er^{3+}/Yb^{3+} co-doped phosphate glasses. *Opt. Mater.* **2018**, *77*, 161–169. [CrossRef]
214. Milenko, K.; Konidakis, I.; Pissadakis, S. Silver iodide phosphate glass microsphere resonator integrated on an optical fiber taper. *Opt. Lett.* **2016**, *41*, 2185–2188. [CrossRef]
215. Dong, C.H.; Yang, Y.; Shen, Y.L.; Zou, C.L.; Sun, F.W.; Ming, H.; Guo, G.C.; Han, Z.F. Observation of microlaser with Er-doped phosphate glass coated microsphere pumped by 780 nm. *Opt. Commun.* **2010**, *283*, 5117–5120. [CrossRef]
216. Mai, H.H.; Kaydashev, V.E.; Tikhomirov, V.K.; Janssens, E.; Shestakov, M.V.; Meledina, M.; Turner, S.; Van Tendeloo, G.; Moshchalkov, V.V.; Lievens, P. Nonlinear Optical Properties of Ag Nanoclusters and Nanoparticles Dispersed in a Glass Host. *J. Phys. Chem. C* **2014**, *118*, 15995–16002. [CrossRef]
217. Adamiv, V.T.; Bolesta, Y.V.; Gamernyk, R.V.; Karbovnyk, I.D.; Kolych, I.I.; Kovalchuk, M.G.; Kushnir, O.O.; Periv, M.V.; Teslyuk, I.M. Nonlinear optical properties of silver nanoparticles prepared in Ag doped borate glasses. *Physica B* **2014**, *449*, 31–35. [CrossRef]
218. Ferrari, P.; Upadhyay, S.; Shestakov, M.V.; Vanbuel, J.; De Roo, B.; Kuang, Y.; Di Vece, M.; Moshchalkov, V.V.; Locquet, J.-P.; Lievens, P.; et al. Wavelength-Dependent Nonlinear Optical Properties of Ag Nanoparticles Dispersed in a Glass Host. *J. Phys. Chem. C* **2017**, *121*, 27580–27589. [CrossRef]
219. Chen, F.; Cheng, J.; Dai, S.; Xu, Z.; Ji, W.; Tan, R.; Zhang, Q. Third-order optical nonlinearity at 800 and 1300 nm in bismuthate glasses doped with silver nanoparticles. *Opt. Express* **2014**, *22*, 13438–13447. [CrossRef]
220. Amjad, R.J.; Sahar, M.R.; Ghoshal, S.K.; Dousti, M.R.; Riaz, S.; Tahir, B.A. Enhanced infrared to visible upconversion emission in Er^{3+} doped phosphate glass: Role of silver nanoparticles. *J. Lumin.* **2012**, *132*, 2714–2718. [CrossRef]
221. Konidakis, I.; Zito, G. Pissadakis, S. Silver plasmon resonance effects in $AgPO_3$/silica photonic bandgap fiber. *Opt. Lett.* **2014**, *39*, 3374–3377. [CrossRef] [PubMed]
222. Schmidt, M.A.; Argyros, A.; Sorin, F. Hybrid Optical Fibers—An Innovative Platform for In-Fiber Photonic Devices. *Adv. Optical Mater.* **2016**, *4*, 13–36. [CrossRef]
223. Jain, C.; Rodrigues, B.P.; Wieduwilt, T.; Kobelke, J.; Wondraczek, L.; Schmidt, M.A. Silver metaphosphate glass wires inside silica fibers—a new approach for hybrid optical fibers. *Opt. Express* **2016**, *24*, 3258–3267. [CrossRef] [PubMed]
224. Ballato, J.; Peacock, A.C. Perspective: Molten core optical fiber fabrication—A route to new materials and applications. *APL Photonics* **2018**, *3*, 120903. [CrossRef]
225. Kang, S.; Dong, G.; Qiu, J.; Yang, Z. Hybrid glass optical fibers-novel fiber materials for optoelectronic application. *Opt. Mater. X* **2020**, *6*, 100051. [CrossRef]
226. Ballato, J.; Dragic, P.D. Glass: The carrier of light—Part II—A brief look into the future of optical fiber. *Int. J. Appl. Glass Sci.* **2021**, *12*, 3–24. [CrossRef]
227. Pisco, M.; Cusano, A. Lab-On-Fiber Technology: A Roadmap toward Multifunctional Plug and Play Platforms. *Sensors* **2020**, *20*, 4705. [CrossRef]
228. Kostovski, G.; White, D.J.; Mitchell, A.; Austin, M.W.; Stoddart, P.R. Nanoimprinted optical fibres: Biotemplated nanostructures for SERS sensing. *Biosens. Bioelectron.* **2009**, *24*, 1531–1535. [CrossRef]
229. Barucci, A.; Cosi, F.; Giannetti, A.; Pelli, S.; Griffini, D.; Insinna, M.; Salvadori, S.; Tiribilli, B.; Righini, G.C. Optical fibre nanotips fabricated by a dynamic chemical etching for sensing applications. *J. Appl. Phys.* **2015**, *117*, 053104. [CrossRef]
230. Hutter, T.; Elliott, S.R.; Mahajan, S. Optical fibre-tip probes for SERS: Numerical study for design considerations. *Opt. Express* **2018**, *26*, 15539–15550. [CrossRef] [PubMed]
231. Lu, F.; Huang, T.; Han, L.; Su, H.; Wang, H.; Liu, M.; Zhang, W.; Wang, X.; Mei, T. Tip-Enhanced Raman Spectroscopy with High-Order Fiber Vector Beam Excitation. *Sensors* **2018**, *18*, 3841. [CrossRef] [PubMed]

applied
sciences

Review

Active and Quantum Integrated Photonic Elements by Ion Exchange in Glass

Giancarlo C. Righini [1,†] and Jesús Liñares [2,*,†]

[1] Microdevices for Photonics Laboratory (MIPLAB), Nello Carrara Institute of Applied Physics (IFAC-CNR), Via Madonna del Piano 10, 50019 Metropolitan City of Florence, Italy; righini@ifac.cnr.it
[2] Quantum Materials and Photonics Research Group, Optics Area, Department of Applied Physics, Faculty of Physics/Faculty of Optics and Optometry, Universidade de Santiago de Compostela, Campus Vida s/n, E-15782 Santiago de Compostela, Galicia, Spain
* Correspondence: suso.linares.beiras@usc.es
† These authors contributed equally to this work.

Abstract: Ion exchange in glass has a long history as a simple and effective technology to produce gradient-index structures and has been largely exploited in industry and in research laboratories. In particular, ion-exchanged waveguide technology has served as an excellent platform for theoretical and experimental studies on integrated optical circuits, with successful applications in optical communications, optical processing and optical sensing. It should not be forgotten that the ion-exchange process can be exploited in crystalline materials, too, and several crucial devices, such as optical modulators and frequency doublers, have been fabricated by ion exchange in lithium niobate. Here, however, we are concerned only with glass material, and a brief review is presented of the main aspects of optical waveguides and passive and active integrated optical elements, as directional couplers, waveguide gratings, integrated optical amplifiers and lasers, all fabricated by ion exchange in glass. Then, some promising research activities on ion-exchanged glass integrated photonic devices, and in particular quantum devices (quantum circuits), are analyzed. An emerging type of passive and/or reconfigurable devices for quantum cryptography or even for specific quantum processing tasks are presently gaining an increasing interest in integrated photonics; accordingly, we propose their implementation by using ion-exchanged glass waveguides, also foreseeing their integration with ion-exchanged glass lasers.

Keywords: ion-exchanged glass; active optical waveguides; quantum integrated optics; integrated photonics

Citation: Righini, G.C.; Liñares, J. Active and Quantum Integrated Photonic Elements by Ion Exchange in Glass. *Appl. Sci.* **2021**, *11*, 5222. https://doi.org/10.3390/app11115222

Academic Editor: Alessandro Belardini

Received: 19 April 2021
Accepted: 28 May 2021
Published: 4 June 2021

Publisher's Note: MDPI stays neutral with regard to jurisdictional claims in published maps and institutional affiliations.

1. Introduction

Ion exchange in glass has been exploited since the early Middle Ages to color (stain) glasses: marvelous stained glass windows, obtained by ion exchange from a silver paste, are still visible in many European cathedrals. The technique of staining glass yellow by painting it with silver salts was already in use in the 14th century: the glass painter Antonio da Pisa, besides creating, in 1395, a superb window in the south door of Florence Cathedral (Figure 1), was the author of a treatise on glass painting, where he shortly described how to produce the yellow color [1]. The glass surface is coated with a silver compound dispersed in a clay medium; during the heating just above the glass transition temperature the silver ions are exchanged with the alkali ions present in the glass (mostly Na^+ or K^+), which then diffuse into it. Subsequently, due to the presence of impurities in the glass, silver ions reduce to metallic silver nanoparticles; the resulting color depends on the size, shape and concentration of the nanoparticles.

For large-scale industrial applications, however, the ion-exchange technology had to wait until the beginning of 20th century, when the chemical surface tempering of glass started to be used [2]. The process (sometimes also called ion stuffing or just chemical strengthening) became a standard in the industry in the 1960s, when, for instance, Corning developed Chemcor® glass, a chemically strengthened glass that was intended to be used in phone booths, prison windows, eyeglasses and automobile windshields. In parallel to these events, there were interesting advances in the optical field, too: in the 1910s it was observed that the ion-exchange process was also inducing a change in the refractive index of the glass, and in the 1960s integrated optics moved its first steps, with waveguiding structures that required materials having slightly different refractive indices [3].

Figure 1. Image of a part of a two-lancet stained window in the Florence Cathedral, representing San Miniato (St Minias) on the left and San Paolo (St Paul) on the right. The artwork was made in 1395 by Antonio da Pisa and is recorded in the Italian Stained Glass Windows Database (BIVI). Reproduced with permission of Opera di Santa Maria del Fiore, Firenze.

Since then, ion exchange has been proven to provide an excellent technological platform for optical integration. Optical waveguides in glass by ion exchange were first fabricated in 1972 using electrically induced migration of thallium ions into a borosilicate crown glass that contained K_2O and Na_2O [4], but soon waveguides were also produced by thermal ion exchange from thallium, potassium or silver nitrate, with Ag^+ giving the best results (besides also being much safer than Tl^+) [5]. Quite obviously, there was also a growing industrial interest, as testified by several patents; for instance, the US Patent 3,857,689 by K. Koizumi et al., filed in 1972 and published in 1974, concerned the "Ion exchange process for manufacturing integrated optical circuits" [6]. Figure 2 shows a schematic view of the process, where the glass plate (11) with the Ti mask (12) is immersed in a molten salt bath (13); as an example, the patent indicated that for a Schott F2 lead silicate glass the process produces a surface refractive index difference $\Delta n = 0.005$ when immersing the glass for about 3 h in a molten salt bath heated at 450 °C and containing Tl_2SO_4 and $ZnSO_4$ at an equal mole ratio [6].

Almost at the same time, the US Patent 3,836,348A by T. Sumimoto et al., filed again by Nippon Selfoc Co. Ltd. in 1973 and published in 1974, presented a "Method for manufacturing optical integrated circuits utilizing an external electric field" [7]. They claimed that, by using Tl^+ ions and an electrical field to drive the ion diffusion, a refractive index profile close to a rectangular one can be achieved.

In the following years, the fabrication technology of glass waveguides employing several ion exchangers (K^+, Cs^+, Rb^+, Cu^+, Li^+ and so on) experienced a great development, and commercial devices were designed and tested (see, for instance, Reference [8]). In parallel, ion-exchange technology was also exploited to create gradient index (GRIN) glass rods [9] and rod lenses with parabolic index profile, suitable for photographic systems [10] and also in conventional optical systems [11] but mostly for laser-to-fiber or fiber-to-fiber coupling [12]. Recently, it was shown that the use of GRIN lens cascades is opening up many surprising applications, spanning from quantum optics to clinical diagnostics [13].

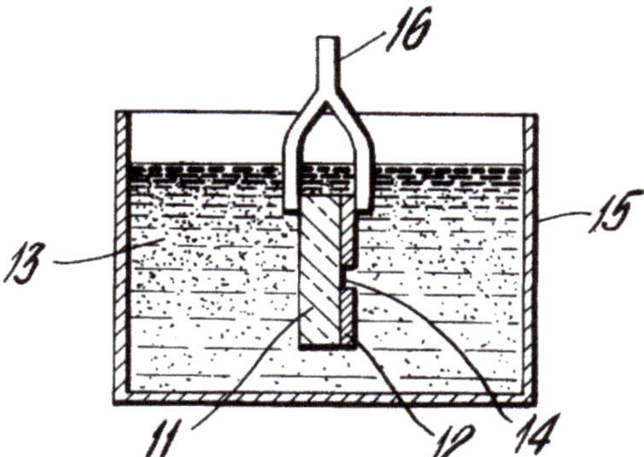

Figure 2. Schematic view of the ion-exchange process to fabricate a gradient-index channel waveguide in a glass substrate [6].

Further theoretical and experimental advances completed the foundations of what is now a fully mature technology for integrated optics and photonics; with time, several very interesting review papers appeared, showing the gradual progresses of this R&D field, up to the use of femtosecond lasers to drive the ion diffusion in glasses [14–21]. Moreover, as ion exchange in glass allows producing either surface or buried waveguides, this technology is promising for 3D integration, too [22,23].

A sketch of the K^+/Na^+ exchange process, which is exploited quite often because it guarantees waveguides with low propagation loss, is shown in Figure 3 [24]. The K^+ ions from the molten salt enter the glass, whereas Na^+ ions from the glass move to the melt, in order to keep the electrical neutrality. This type of exchange has two drawbacks: the induced refractive index variation Δn is lower than 10^{-2} (so, only a weak guide confinement is possible), and it introduces some birifringence, due to the compressive stress induced by the difference in ionic radius between Na^+ and K^+ ions. On the other hand, as already mentioned, this stress may also be exploited to increase the mechanical resistance of the glass itself (chemical strengthening).

The Ag^+/Na^+ ion exchange, instead, thanks to the similar size and the difference of polarizability of the two ions, allows a higher Δn (10^{-1}) without birefringence and also the burial of waveguides in a two-step process, as sketched in Figure 4. Unfortunately, silver, too, has some limits: in the presence of humidity, surface waveguides (which are used in sensing devices, interacting with external medium) suffer higher losses, due to chemical reduction to metal silver. An approach useful to overcome this limitation is to use the Tl/Na process [25]: in this case, a drawback is due to the toxicity of thallium. Thus, we must conclude that the choice of the ion pair to be used in the waveguide fabrication by ion exchange is dictated by the characteristic (optical, mechanical, safety) one wants to privilege.

In this paper, the focus is on glass ion-exchanged integrated optical devices: their applications cover broad areas, where optical sensing and optical communications are at the forefront. Thus, in the field of optical sensing, many integrated optical sensors are based on the properties of the evanescent modal optical field, and ion-exchanged waveguides offer the advantage of the proximity between the glass index ($n = 1.5$) and the index of frequently used organic compounds (around 1.4–1.6) and of aqueous solutions (index close to 1.33). These guided-wave structures lead to large evanescent fields, which maximize the optical sensitivity function. A large amount of optical integrated sensors and biosensors based on ion-exchanged glass waveguides have been proposed [18–20,26] and keep being implemented in the last years [27–30]. An interesting example may be represented by a 32-analyte integrated optical fluorescence-based multi-channel sensor and its integration to an automated biosensing system [31]. Figure 5 shows the layout of the fiber-pigtailed sensing chip consisting of a single-mode channel waveguide circuit that distributes the light to 32 separate sensing patches on the chip surface. The use of this waveguide immunosensor, based on a K^+-exchanged BK7 glass, was demonstrated for the detection of Microcystin-LR cyanotoxin in two real lake water samples taken from Fuhai lake and Beihai lake in Beijing [31].

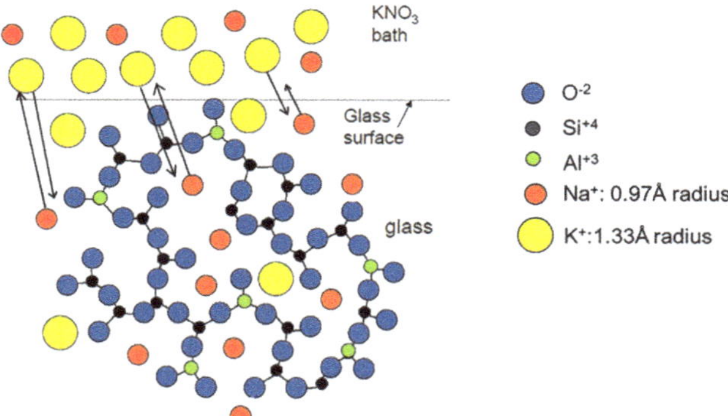

Figure 3. Depiction of the K^+/Na^+ ion-exchange process using KNO_3 molten salt as the source of K^+ ions. The glass considered here is an aluminosilicate glass with an alkali oxide (Na_2O) as glass modifier. The kinetics of the process is described by diffusion equations. Reproduced from [24] with permission by John Wiley and Sons.

Surface waveguide

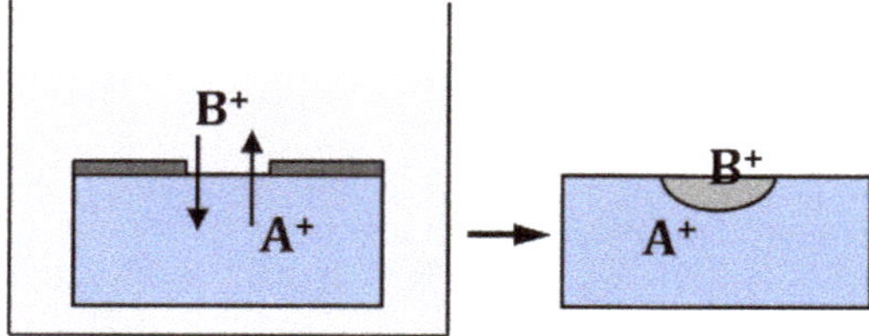

Buried waveguide

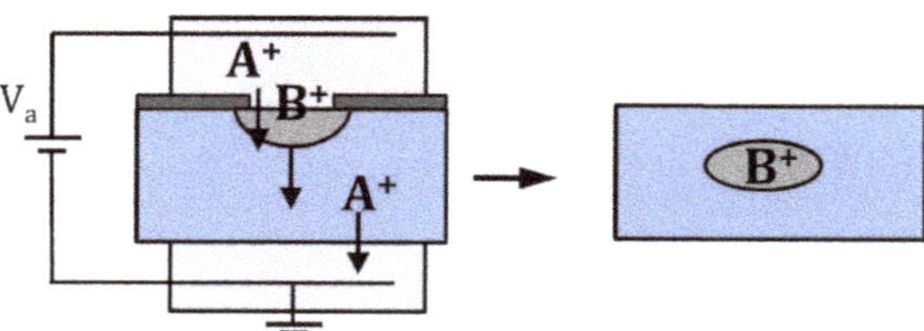

Figure 4. Outline of two different ion-exchange processes involving two ions A and B (e.g., K^+ and Na^+, respectively): the upper drawing refers to a purely thermal diffusion, that produces a surface waveguide. The bottom drawing indicates a field-assisted diffusion, that leads to a buried waveguide. Reproduced with modifications from [25] with permission by Elsevier.

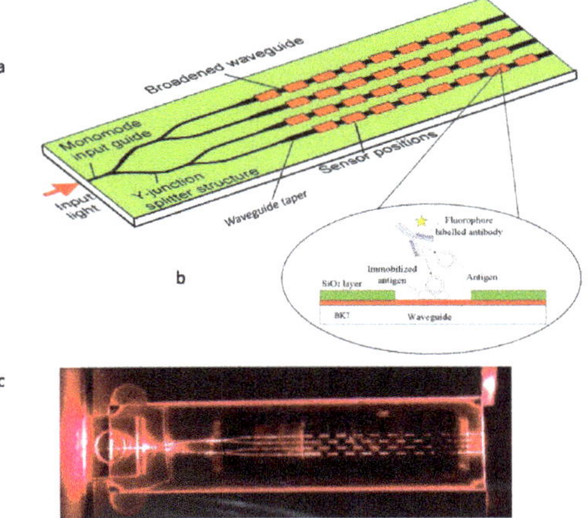

Figure 5. (**a**) Integrated optical sensor's layout; (**b**) cross-sectional view, showing waveguide, isolation layer and location of the surface chemistry; (**c**) photo of light propagation along the waveguide chip (Photo credit: Lanhua Liu). Reproduced from [31] under a Creative Commons CCL4.

In the field of optical communications, the high optical compatibility between glass integrated devices and optical fibers clearly offers an important advantage with respect to other materials. Again, ion-exchanged glass waveguides have constituted a fundamental component of several devices based on modal coupling, such as directional couplers (e.g., for power dividers), gratings (e.g., for contradirectional coupling in integrated lasers), multimode interferometers (MMI), $1 \times N$ and $2 \times N$ splitters, array waveguide gratings (e.g., for wavelength division multiplexing), and so on [19,32,33]. Section 2 summarizes the fundamentals of ion-exchanged channel waveguides and circuits. By using suitable glass substrates, such as rare-earth doped oxide glasses, active integrated optical devices may also be fabricated through ion exchange; this topic is discussed in more detail in Section 3 of this paper. Other interesting aspects concern the manufacturing of nonlinear optical (NLO) integrated devices: Jackel et al. [34], in 1990, designed two glasses, containing Ti and Nb to increase the nonlinear index coefficient and an alkali ion (Na^+ or K^+), so to have a large optical nonlinearity and to be compatible with ion exchange. Spatial soliton propagation in these waveguides was demonstrated, which proved the possibility of implementing an all-optical switch. Almost at the same time, tests were also performed on commercial semiconductor-doped glasses, which had been proven to possess nonlinear optical properties [35,36]: optical waveguides were fabricated by using a $Cs^+ - K^+$ exchange [37]. Ion-exchanged waveguides may also be capable of handling high laser powers, e.g., from an integrated Q-switched laser [38], and are therefore suitable, due their intrinsic light confinement properties, for nonlinear applications [39]. In that area, however, hybrid structures may be more efficient and the coupling of a lithium niobate film with a ion-exchanged glass waveguides can lead to the manufacturing of electro-optic modulators or second harmonic generators [40].

Many other applications are possible, also linked to the new topics emerged in the last years, as for example integrated quantum photonics (IQP) for the implementation of tasks in quantum information, particularly, in quantum communications. It has been underlined that in IQP 'the big challenge is how to realize a scalable, convenient platform for practical deployment and commercialization' [41]. Indeed, the growth of this area requires the design and fabrication of passive and active integrated devices to be used, for example, in quantum photonic simulation, quantum cryptography, quantum photonic sensors and so on. It is well known that, since the early experiments on silica-on-silicon structures [42], different platforms have been proposed for the implementation of IQP, including those ones silicon-based, III-V semiconductors and lithium niobate [43–46]. It is very difficult, however, for a single material to meet the stringent demands of most quantum applications, so that hybrid platforms combining different technologies in a single functional unit may have greater success [47]. Silica-based platforms have been considered, too, but only with reference to femtosecond laser written circuits [44,48]. To our knowledge, despite the suggestion, several years ago, by the group of one of the present authors [49], a platform based on ion-exchanged glass (IExG) circuits has not been considered so far as an effective integration technology for quantum photonics. This topic is discussed in more detail in Section 4, where we present a brief review of the most incipient proposals and results in the field of quantum circuits based on an IExG platform. Likewise, it is shown that with hybrid glasses active and quantum integrated devices can be combined in an efficient way by using IExG platform, in particular, IQP devices for generating and detecting quantum states are presented; hybrid integration with lithium niobate films may also permit the realization of efficient nonlinear integrated devices [40]. Finally, we present in Section 4 the design, fabrication and preliminary experimental results of a quantum projector for detecting quantum states by projective measurements, for example for quantum cryptography through multicore optical fibers (MCF) [50,51] or even few-mode optical fibers (FMF) [52]; the primary aim is to make clear the potentiality of the IExG platform for integrated quantum photonics.

2. Fundamentals of Ion-Exchanged Glass Waveguide Circuits

There are excellents works on ion-exchanged glass waveguide circuits, but here we focus our attention on circuits formed by elements as channel waveguides, directional couplers and optical gratings (contradirectional couplers), and intended for classical and quantum optical sensing and communications. Channel waveguides are the pillar of integrated optics [17–19,53,54] and optical integrated gratings are also fundamental for the implementation of many devices, including integrated amplifiers and lasers in rare-earth doped glasses [19,55]. Directional couplers and $1 \times N$ splitters, too, are based on channel waveguides and have been mainly intended for power derivation in optical fiber communication networks, but they also play an important role in optical sensing and in interferometry [31,56], including applications to optical astronomy [57]. The fabrication of these devices in a glass substrate usually requires a lithographic process [18,19] in order to allow a spatially selective ion-exchange. Let us refer to thermal ion exchange: first, a metallic film (Al or Ti) is deposited on the glass substrate, followed by the deposition of a photoresist film. After UV exposure through a given mask and a subsequent chemical etching, the mask pattern is transferred to the metal film, so that, when immersing the substrate in a molten salt, the ion exchange occurs only in the volume under the designed opening. This process produces surface channel guides, which can be slightly buried by a second thermal diffusion process or strongly buried by using an electrical field.

In Figure 6, the design is shown of an integrated optical device of arbitrary dimension N, that is, N optical paths (channel waveguides) coupled by concatenated directional couplers (DC) and phase shifters (PS). It represents a generic optical device able to implement different passive quantum operations. Active functions, such as optical generation and amplification, may also be added if hybrid glasses, with a portion doped by rare earth elements (on the left in Figure 6), are used.

We must stress that hybrid glasses can be obtained by joining a block of a passive glass with a block of an active glass by a low temperature bonding technique [58,59]; alternatively, the hybrid glass can be also implemented by bonding an active glass superstrate to a passive glass substrate [60]. Consequently, integrated lasers having optical gratings (OG) resonators can also be included in the integrated circuit, as sketched in Figure 6; they are pumped by an external signal P_j $(j = 1, \ldots, N)$, which can generate classical and/or quantum light states [61]. The optical gratings can be fabricated by laser writing or by the same photolithographic procedure described above [17–19]. In short, Figure 6 shows a general ion-exchanged hybrid glass integrated optical circuit highly intended for classical/quantum communications (or even classical/quantum optical sensing) and that can be adapted for implementing different devices as wavelength multiplexers, quantum states generators, quantum states projectors and so on, as is shown in some detail in Sections 3 and 4.

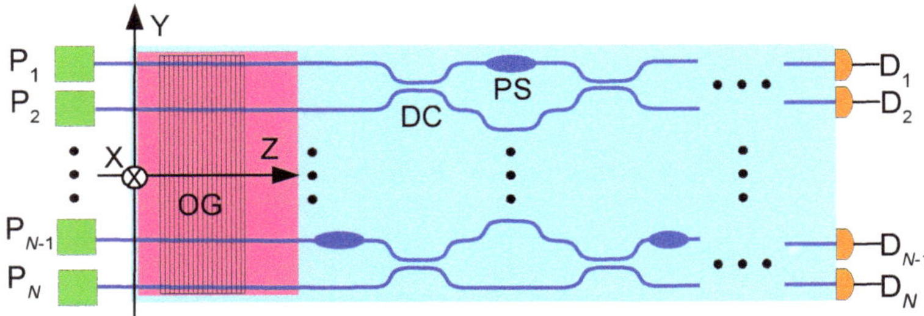

Figure 6. A general device formed with N paths (channel waveguides) in active (**left**) and passive (**right**) sections of a hybrid glass substrate. An optical grating (OG) in the rare-earth-doped glass section (the left part of the substrate) is used to implement a laser structure which, pumped by beams P_j, $(j = 1, \ldots, N)$, generates N laser signals; several directional couplers (DC) and phase shifters (PS) in the passive section perform linear optical transformations. N detectors D_j are also shown.

The total refractive index profile of the above ion-exchanged hybrid glass integrated optical circuit can be represented, in a good approximation, as follows

$$n(x, y, z) = n_s + \sum_j \Delta n_j(x, y, z; p_{j1}, \ldots, p_{jq}) \tag{1}$$

where n_s is the substrate index, and $\Delta n_j(x, y, z; p_{j1}, \ldots, p_{jq})$, $j = 1, \ldots, N$ are functions describing the refractive index profile obtained in each ion-exchanged channel waveguide with parameters $p_{j1}, \ldots, p_{jq}$, that is, ion-exchanged parameters as effective depth d, maximum refractive index change $\Delta n, \ldots$, geometrical lithographic parameters as width w of a channel guide, separation s between channel guides, ... and so on. The centers of the channel guides (usually the position of the maximum refractive index) follow paths described by functions $f_j(z)$, that is, z-functions representing translations of a refractive index profile. A typical case is a channel guide j, which is bent along S-shaped transitions to couple it with other parallel channel guide $j + 1$. These transitions may follow, for example, the path described by Minford et al. (Minford curve) for obtaining low losses [62]. An example of a refractive index profile representation for an arbitrary channel guide j is the following (from here and for the sake of simplicity we omit subscript j)

$$\Delta n(x, y, z) = \Delta n \operatorname{Erfc}\left(\frac{x}{d_{ox}}\right) \exp\left\{-\frac{(y - f(z))^2}{d_{oy}^2}\right\} \tag{2}$$

where Erfc is the complementary error function, which is a typical solution for linear ion exchange, although more general solutions can be chosen [17,63–65], $f(z)$ is the curve (path) profile, Δn is the maximum refractive index change and d_{ox} and d_{oy} are effective depths along x and y direction. Note that we have assumed a factorizable index profile, $\Delta n(x, y, z) = \Delta n\, E(x) G(y, z)$, that is, a product between an index profile in depth (x-direction) and a Gaussian index profile along y and z directions describing quite well the lateral diffusion. In many cases such a factorization is a good approximation, or even it can be forced for the sake of design simplicity [66]. Moreover, it is well known that there are different approximate methods to be used in order to obtain manageable mode solutions such as, for example, the effective index method (EIM) [16,66], perturbative and variational methods [16] and so on. On the other hand, this kind of function can also describe the change of index or geometry for other integrated optical elements as phase shifters, for example, those based on an adiabatic approximation that could be represented by the following function

$$\Delta n(x, y, z) = \Delta n \, \mathrm{Erfc}(x/d_{ox}) \, \exp\{-\frac{y^2}{d_y^2(z)}\} \tag{3}$$

Note that in this case we have a rectilinear channel waveguide ($f(z) = 0$) with $d_y(z)$ an adiabatic z-function and therefore the effective index changes slowly on z and thus a well defined phase shifting can be obtained.

We must recall that the cornerstone of the ion-exchange method is that these functions are, in a good approximation, proportional to the concentration $C_j(x, y, z)$ of exchanged ions. In fact, the q parameters $p_1, \ldots, p_q$ such as maximum refractive index change Δn, effective depth d_{ox}, lateral effective depth d_{oy} and so on, can be expressed in an empiric way as functions of both the fabrication physical parameters, such as temperature T, ionic concentration $[C]$ of the molten salt (or ionic concentrations $[C_1, C_2, \ldots]$, as in the case of molten salts eutectic mixtures), ion-exchange time t and also geometrical parameters, as width of the channel guide mask w, separation between channel guides s and so on. We must also underline that, although an empiric approach can be sufficient in many cases to determine the fabrication parameters, a great amount of research work has been devoted to a rigorous physical modeling and numerical simulation of ion exchange in glass, which in turn, is related to the development of different theories on ion exchange in glass [15,67–71]. These physical/numerical approaches can be required when the refractive index profile must be known with a great accuracy. In any case, one of the main aims in integrated optical circuits is to obtain single-mode channel guides (paths) that are coupled in certain regions by directional coupling and with phase shifters suitably located, as shown in Figure 6 to achieve an integrated optical circuit with a specific purpose. Finally, it is also worth indicating that sometimes two-mode or even few-mode waveguides can be considered. The modes can be considered in plane yz or also in depth x. In the latter case, it can be considered as a type of three-dimensional integration [27,72]. On the other hand, two-mode guides in the yz plane have also been considered, for example, for spatial multiplexing [73].

3. Ion-Exchanged Active Integrated Photonic Devices

As already briefly mentioned in the previous sections, ion exchange may occur in a variety of substrates and using many different ions. The process, however, can be easily performed especially when using univalent ions and glasses whose composition includes alkali ions as glass modifiers. Similarly, in crystalline materials such as lithium niobate, only Li^+ ions are effective and they exchange with H^+ ions. These limitations, however, are not very critical, as testified by the large number of photonic devices that have been successfully implemented. Among those devices, a very important class is constituted by the active ones, i.e., integrated optical frequency converters, amplifiers and lasers. Let us here briefly overview some of the excellent results already achieved, limiting ourselves to devices produced by ion exchange in rare-earth-doped (RED) glasses.

It has already been mentioned that RED glasses constitute an excellent physical medium for generating and amplifying light. Indeed, oxide glasses are well-known as excellent hosts for rare-earth ions: not by chance, one of the very first solid-state lasers was demonstrated in 1961 by Snitzer in American Optical Company using a Nd^{3+}-doped glass [74]. In that same year Snitzer and colleagues constructed and operated the world's first optical fiber laser and three years later the first optical fiber amplifier [75]. Even the first thin-film waveguide glass amplifier, in 1972 [76], and the first integrated optical glass laser, in 1974 [77] were implemented in the same material, Nd^{3+}-doped glass and operated at ~1.06 μm wavelength. It has to be noted that, whereas the amplifier was fabricated by RF-sputtering deposition of the doped guiding film, the waveguide laser was indeed realized in a Nd-doped borosilicate glass by using the double-diffusion process with an electric field already reported by Izawa and Nagakome [4]. The channel waveguides had an almost circular profile, and the pump threshold was around 18 µJ, namely about one half of the threshold of the corresponding bulk laser [77].

The excellent properties of ion-exchanged waveguide lasers had therefore been demonstrated, but a strong interest in these devices emerged only much later, after the implementation, around 1985, of the second generation of fiber communication systems, which used single-mode fibers. Thus, for instance, in 1989 Najafi et al. studied the fabrication of optical waveguides in commercial Kigre Nd-doped lithium-silicate glass by means of Ag-Li exchange and verified that the process did not influence the emission properties of Nd [78]. In 1990, a waveguide laser emitting at 1054 nm was reported by Aoki et al. [79]; the laser, end-fire pumped by a diode-laser operating at 802 nm, was capable of a maximum output power higher than 150 mW. It was fabricated in a Nd^{3+}-doped glass (Hoya LHG-5 phosphate glass, 3.3 wt% Nd) by using field-enhanced silver ion exchange. The same group also demonstrated a gain of 3.4 dB at 1.054 µm and an efficiency of 0.027 dB/mW in a 10 mm long waveguide amplifier pumped with an argon-ion laser [80]. Several other works were performed in the early 1990s concerning Nd-doped waveguide lasers and amplifiers, still using ion-exchange fabrication technology [81–85]; it was also proven that distributed Bragg reflector (DBR) integrated lasers could be fabricated by realizing holographically an OG in photoresist and then etching it into the channel guide by argon ion milling [83]. By using ion exchange in a semiconductor microcrystallite-doped glass second harmonic generation of Nd:YAG laser radiation was also achieved [36].

In the meantime, the evolution of fiber communication systems continued and saw the introduction of 1.5 µm laser diode sources, owing to the fact that silica fibers have the lowest propagation losses in that wavelength window. Fiber lasers and amplifiers had therefore to use another rare earth than Nd, with photoluminescence emission in the near infrared band: this was Erbium, which is characterized by a strong emission at 1.532 µm. Er-doped fiber amplifiers (EDFAs) soon became a very important component of fiber transmission systems, that needed regeneration and re-amplification of the guided light after a certain distance, of the order of tens of kilometers. Most of the pioneering research on EDFAs was performed at Southampton University by the David Payne's group and at AT&T Bell Labs by Emmanuel Desurvire and colleagues: the two groups published the first successful results almost at the same time, in 1987 [86,87]. A strong limitation to the practical use of these early fiber amplifiers was related to the pump signal: Payne's group was using pump wavelengths in the range 655–675 nm, whereas Desurvire used a pump at 514.5 nm. In both cases, bulk pump lasers were necessary. A substantial step forward market diffusion of EDFAs was made with the availability of powerful laser diodes. Then, the demand of ever increasing system transmission capacity and of a mechanism for creating multipoint networks led to the use of the wavelength division multiplexing (WDM) technology. As integrated optical devices could offer a cheap and small-footprint approach to the fabrication of WDM devices, a renewed R&D interest was focused on Er-doped glass devices and to the integration of passive and active functions on a same glass substrate. An issue, however, was related to material requirements, much more stringent for integrated optics than for fibers, due to the different fabrication technologies and to the much higher rare-earth concentration required in short-length planar devices. An interesting effort to assess the relative merits of different compositions of the host glass was published by Miniscalco in 1991 [88]. He concluded that 'in most situations' Al-doped silica was the preferred glass for Er^{3+}-doped fiber amplifiers at 1500 nm. In planar short-length devices, silicate glasses may suffer even more from the lower solubility of rare-earth-ions with respect to phosphate glasses, since it may lead to clustering and reducing the conversion efficiency. In spite of this, the first ion-exchanged waveguide laser emitting at 1540 nm used a silicate BK7 glass containing 0.5 wt% Er_2O_3 [89]. The laser cavity was formed by bonding dielectric mirrors to the chip end facets, and a thermal ion exchange in a KNO_3 bath was employed in waveguide fabrication. In the same year 1992, an Er-doped waveguide amplifier with 0.65 dB/cm net gain with 100 mW pump power was demonstrated [90]. This device was constituted by a phosphorus-doped silica core doped with 0.55% wt Er and silica cladding deposited by flame hydrolysis on a silicon substrate. Quite obviously, owing to the appeal of erbium-doped waveguide amplifiers (EDWAs), various technological approaches to waveguide fabrication were pursued, with deposition techniques of doped glass films (e.g. RF-sputtering, sol–gel, PECVD, flame hydrolysis) competing with diffusion techniques (ion exchange, ion implantation) in bulk doped glasses [20]. As an example of RF-sputtered amplifying waveguides, a net optical gain of 4.1 dB/cm at 1.535 µm was achieved by Polman et al. by depositing Er-doped multicomponent phosphate glass films [91]; the same group also reported about the possibility of using MeV ion implantation to introduce erbium in a variety of thin films, from oxide glasses to ceramics and to amorphous and crystalline silicon [92].

A commercial interest was also emerging. Among the glass suppliers, the Corning company was particularly interested in the development of ion-exchangeable Er-doped glasses and of amplifiers, as testified also by two patents [93,94]; in the latter one, an optimized composition of a boron-free silicate glass that could be doped with up to 5% wt erbium oxide and ion exchanged with thallium is described [94]. More waveguides and devices based on Er-doped glass substrates and ion-exchange technology were reported in the following years [90,95–107]; besides waveguide lasers and amplifiers, optical upconversion devices were investigated, too [101]. At a certain point, phosphate glasses seemed to allow higher performances to ion-exchanged EDWAs, as they could be doped with larger amounts of rare-earth ions than silicates. Even in phosphate glasses, however, high erbium concentrations may produce quenching of the luminescence; moreover, the absorption cross section of Er^+ ions at ~980 nm is small. Thus, the solution was to add another rare-earth ion as a sensitizer for erbium. Glass co-doping with ytterbium ions has proven to be a very efficient solution: it allows the transfer of energy from excited Yb^{3+} ions to close Er^{3+} ions through a cooperative cross-relaxation process, with the result of significantly enhancing system absorption at 980 nm and making the pumping mechanism more efficient. Moreover, in the case of high concentration, the Er^{3+} ions are closer together, so that deleterious non-radiative energy exchanges between neighboring ions can take place; the presence of Yb^{3+} ions is also effective in reducing such unwanted Er^{3+} - Er^{3+} ion energy transfer interactions by increasing the mean inter-atomic distance. A useful spectroscopic investigation of an extensive series of Er^{3+}-doped and Er^{3+}-Yb^{3+}-co-doped soda–lime–silicate (SL) and aluminosilicate (AS) glasses was also presented in a paper by Hehlen et al. [108]. AS glasses showed higher oscillator strengths and larger inhomogeneous broadening of $4f$ transitions than SL glasses; it was also suggested that a minimum molar concentration ratio 2:1 of Yb oxide to Er oxide is required to achieve efficient sensitization of Er^{3+}. In 1995, an integrated laser was produced by K^+/Na^+ ion exchange in an alkali-rich barium silicate glass containing 17% wt Y_2O_3 and 1.5% wt Er_2O_3 [109]. Later, commercial Er-Yb co-doped phosphate glasses became available, too, where single-mode channel waveguides could be produced by K^+/Na^+ ion exchange; a net gain of 7.3 dB in a 6 mm long waveguide amplifier was claimed [110].

EDWA devices with excellent characteristics were obtained in both silicate and phosphate glasses co-doped with Er and Yb; as an example, a lossless 1×2 splitter was demonstrated in a buried thallium-exchanged waveguide in a borosilicate glass co-doped with 5% wt Yb_2O_3 and 3% wt Er_2O_3 [111]. They measured 2.3 dB/cm gain and 0.07 dB/mW gain efficiency in a 3.9 cm long straight waveguide amplifier with a pump of 130 mW. Almost at the same time, Barbier et al. achieved a 7 dB net gain in a 41 mm long amplifier with double-pass configuration ($\approx$1.7 dB/cm) using a buried waveguide produced by two-step ion exchange in a phosphate glass doped with 2.5 % by weight of erbium and 3% by weight of ytterbium [112]. Shortly thereafter, however, the same group reported several impressive advances in EDWA's implementation, including: an amplifying four-wavelength combiner, consisting of an all-connectorized 4×1 glass splitter followed by a 4.5-cm-long Er/Yb-doped waveguide amplifier ($\approx$2.6 dB/cm) [98]; a planar amplifier capable of 3.1 dB/cm net gain [113]; a multiplexer/amplifier gain-block constituted by an amplifying section realized in phosphate glass as a 5.5 cm straight waveguide planar waveguide amplifier and a 980/1550 nm multiplexer realized as a 3 cm long directional coupler in a silicate glass [114]. The amplifier and the multiplexer were assembled together via active alignment followed by bonding with UV curing glue; the module had a fiber-to-fiber gain of 15.7 dB and a noise figure of 4.7 dB at 1535 nm. In all these cases, the amplifier was realized in a phosphate glass co-doped with 2 wt.% erbium and 4 wt.% ytterbium.

In the 2000s, the literature was enriched by several publications in this area, which can be classified into three groups:

(a) Papers dealing with the synthesis and characterization of rare-earth doped glasses suitable for ion-exchange fabrication of active devices [17,115–144].
(b) Demonstration and characterization of rare-earth doped ion-exchanged waveguide amplifiers and frequency converters [140,145–165].
(c) Demonstration and characterization of rare-earth doped ion-exchanged waveguide lasers [60,61,147,166–169].

An example of a laser device is shown in Figure 7: it is a 3D hybrid structure, with a channel core waveguide and a slab core waveguide which, in a certain volume, interact with each other and with the Bragg grating that provides the optical feedback necessary for laser action. Figure 8 shows the longitudinal and cross sections of the device. The passive glass is a GO14 silicate by Teem Photonics, whereas the active glass is a IOG-1 phosphate by Schott, doped with 2.2 wt% Er_2O_3 and 3.6 wt% Yb_2O_3. In both glasses, the waveguides are produced by Ag^+-Na^+ ion exchange.

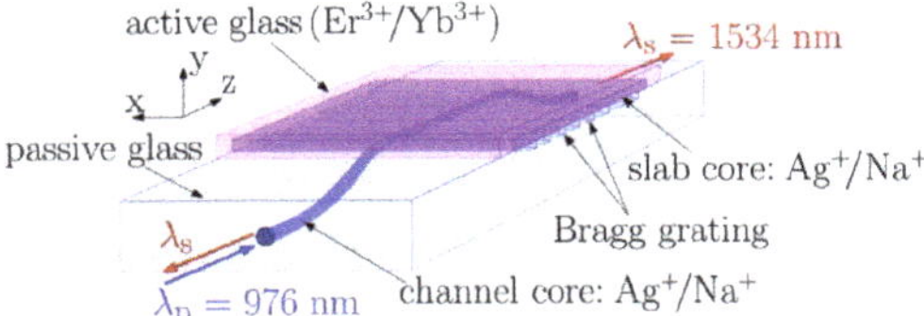

Figure 7. Sketch of a hybrid distributed feedback (DFB) laser constituted by a channel core waveguide, buried in the silicate substrate by and loaded by a slab core waveguide in an Er/Yb-co-doped glass. Both waveguides are made by ion exchange. Reproduced with permission of Elsevier from [60].

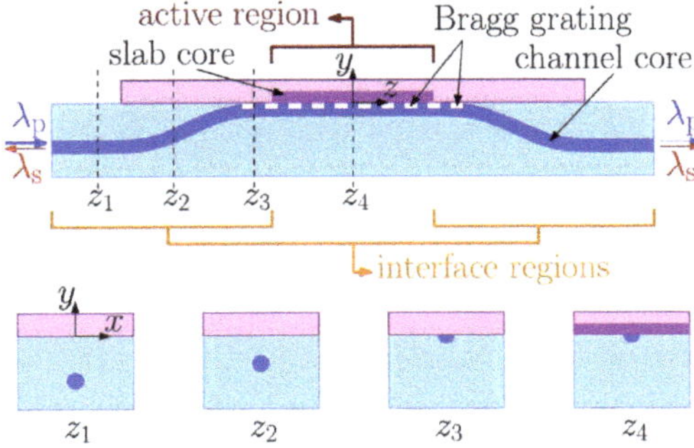

Figure 8. Longitudinal (**top**) and cross (**bottom**) sections of the hybrid DFB laser structure. Reproduced with permission of Elsevier from [60].

An example of a packaged Er-Yb doped waveguide amplifier (EYDWA) is shown in Figure 9 [155]. The substrates were commercially available Schott IOG-1 phosphate glasses co-doped with 2 wt% Er_2O_3 and 5 wt% Yb_2O_3. Channel waveguides were fabricated by ion exchange using an Al thin film mask and were buried by using a two-step process, namely, a K^+-Na^+ thermal exchange ($T = 385\,°C$, $t = 60$ min), followed by a field-assisted annealing ($T = 380\,°C$, $t = 30$ min, electric field 120 V/mm) [123]. The figure shows a photo of an EYDWA device packaged by using fiber silicon V-groove arrays and a UV curing glue. In a pigtailed 4.0 cm long EYDWA, a net gain of 8.0 dB and an average noise figure of 5.1 dB were obtained in the 1530–1560 nm band.

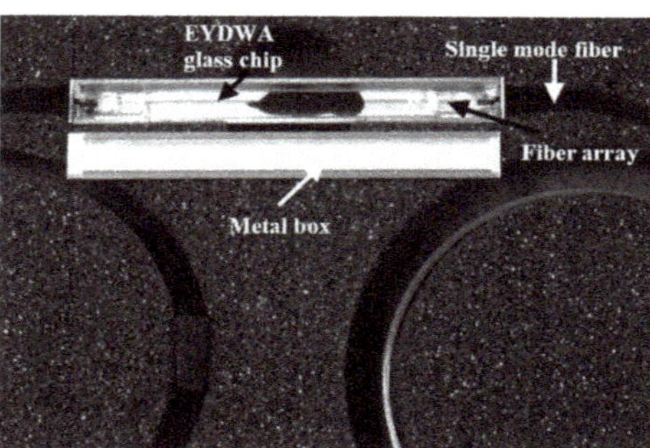

Figure 9. Photo of a packaged EYDWA device. Standard single mode fibers are pigtailed to both the input and output ends of the channel waveguide. The glass chip is then mounted to a metal box to provide mechanical stability and easy handling. Reproduced from [155] with permission by Elsevier.

New research topics have appeared in those years; some attention, for instance, was focused on the possibility of using ion-exchange technology in rare-earth doped tellurite glasses [121,131–133,136,170–172]. Tellurite glasses, i.e., glasses based on tellurium oxide, doped with rare earth have been long studied since they have the lowest phonon energy (about 780 cm^{-1}) among oxide glass formers and, as a consequence, their radiative quantum efficiency is high. Furthermore, they offer good stability, chemical durability, high solubility of rare-earth ions and exhibit high refractive index and a wide transmission range (0.35–6 μm); these properties make them an excellent host material for amplifiers and lasers to be used in a wider wavelength range, e.g., in S + C + L (1460–1615 nm) or C + L + U (1530–1675 nm) bands of the optical communication systems. As already mentioned, other fabrication technologies were also developed, and sol–gel, for instance, has been a very convenient platform for the fabrication of high quality thin glass films, optical planar waveguides and photonic bandgap structures [173–176]. An interesting research concerned the possibility of integrating the sol–gel and ion-exchange techniques for the fabrication of passive and active channel waveguides in rare-earth doped integrated optic devices [115,177].

Among the various research lines aiming at improving the characteristics of waveguide optical amplifiers, it is worth mentioning the article by Donzella et al. [160], where another 3D hybrid structure is designed with the goal of making possible the use of a low-cost high-power broad area laser diode, instead of an expensive single-mode 980 nm laser, for the amplifier's pumping. Their design, sketched in Figure 10 includes direct butt-coupling of the pump light into a multimode passive waveguide (in silicate glass): an Er-Yb doped single mode waveguide (in phosphate glass) is placed on top of the passive one, and light is progressively coupled from bottom to top waveguide, being captured by Yb ions acting as sensitizers for Er ions. Experimental tests with ion-exchanged waveguides fabricated in the two glasses allowed to measure amplified spontaneous emission (ASE); Figure 11 shows the upconversion green light in the upper single mode waveguide, a proof of the effectiveness of the pumping scheme. According to their numerical simulation, gain values of over 3 dB/cm should be achievable.

As already mentioned in Section 2, too, another way of fabricating hybrid active/passive glass structures is to butt join two glass pieces, an undoped one and a rare-earth-doped one, in order to optimize the optical characteristics and minimizing losses. Such a structure was used especially to fabricate lossless splitters and combiners, widely used in telecom systems; to mention some examples, Jaouen et al. separately designed and optimized a passive section in silicate glass and an active section in phosphate glass before joining them by using a UV curing glue, so to produce a lossless 1x8 splitter/combiner [178]. Conzone et al. showed that it was possible to join two aluminophosphate glasses (undoped and doped, respectively) by sandwiching an aqueous phosphate solution between the polished and accurately cleaned surfaces of the two blocks. A rigid joint is formed in 24 h at room temperature. A thermal treatment for 5 h at 375 °C provides a greater chemical and mechanical robustness [58]. Further, Chen et al. reported a low-cost technique for industrial mass production, which was based on vacuum hot press assisted direct bonding applied to custom made multicomponent silicate glasses [179]. A thermal treatment at 605 ± 20 °C for 240 min under a pressure of 28 kPa was sufficient to achieve a very good bonding. Figure 12 shows the first bonded sample. The two glasses were germanoborosilicates with sodium oxide for the ion-exchange process; the doped glass contained 0.5 mol% Er_2O_3 and 1.3 mol% Yb_2O_3.

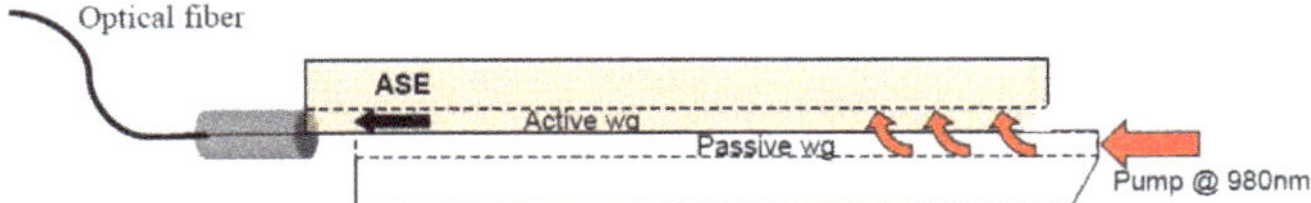

Figure 10. Longitudinal view of the hybrid waveguide amplifier test structure, constituted by a multimode passive waveguide (**bottom**) and a single mode active Er/Yb doped waveguide (**top**). A large area laser diode pump is coupled to the multimode waveguide, and the propagating beam is progressively coupled to the active layer. Energy is absorbed by Yb ions and then transferred to Er ions. Reproduced from [160] with permission by Optical Society of America.

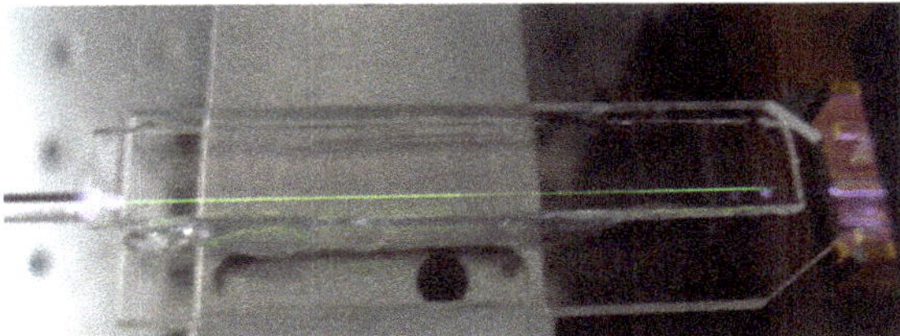

Figure 11. Photo of the green light propagating in the single mode (**upper**) waveguide in the structure of Figure 11. Pump wavelength is around 980 nm, and the green color, due to upconversion emission from excited Er^+ ions, proves that the transfer from the pump to the active layer is effective.

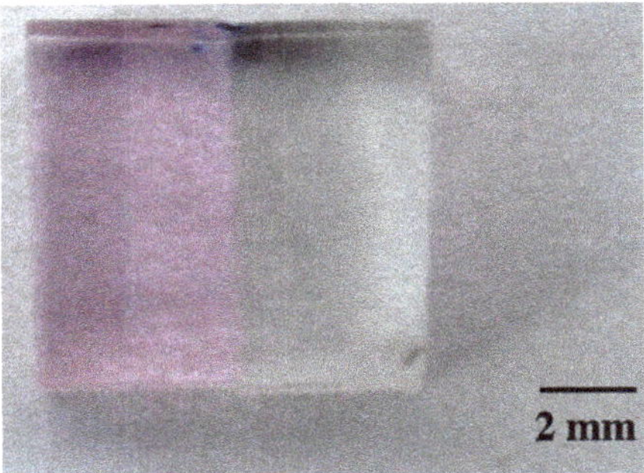

Figure 12. Photograph of the bonded glass sample, with the active (**pink**) and passive (**transparent**) parts. Reproduced from [179] with permission by Elsevier.

A broad review on Er-Yb laser and amplifiers, covering both the photoluminescence issues and the various materials and technologies, was published in 2011 by Bradley and

Pollnau [180]. In the last years, owing to the fact that Er-Yb co-doped ion-exchanged waveguides and devices in silicate and phosphate glasses constitute a mature technology, only very few papers have been published, including some ones related to up–down frequency conversion with application to lighting and solar cells [164,181–184]. The main research focus has been shifted toward technological platforms that could allow integration with silicon and compatibility with CMOS technology; thus, rare-earth doped aluminum oxide has emerged as a new excellent material. Al_2O_3 can be deposited by atomic layer deposition or reactive magnetron sputtering, is transparent from the UV to mid IR and exhibits very low propagation losses; the high solubility for rare-earth ions allows it to provide active functions, amplification and lasing, to silicon and silicon nitride platforms. High-gain erbium amplifiers and erbium, thulium and holmium integrated lasers (with emission around 1.5 µm, 1.8 µm and beyond 2.0 µm, respectively) have been demonstrated [185–187].

4. Quantum Integrated Photonic Devices

As indicated in the Introduction section, different platforms have been proposed for implementing IQP devices. The main platforms are based on III-IV semiconductor, such as gallium arsenide and indium phosphide, lithium niobate and those ones based on silica such as silica-on-silicon, silicon-on-insulator and silicon nitride, as reviewed, for instance, in References [44–46]. IExG platform is not usually included as a possible quantum integrated photonics platform, despite the fact that it is present in many classical optical communications systems and optical sensors. We must indicate that the closest platform to the IExG one is obviously, silica-on-silicon; however, in our opinion, IExG presents a lower technological complexity and retains all the classical advantages of this platform, that is, low-cost, low propagation losses and high compatibility with optical fibers, together with a slightly better modal transverse coupling due to the graded-index profile unlike the strictly step profile of silica-on-silicon waveguides. We must stress that the main limitation, shared with silica-on-silicon, is the low contrast index (therefore low curvature radii and consequently limited device densities). The main consequence is that IExG platform becomes remarkably efficient when quantum applications require only a few qubits. This requirement is fulfilled in most devices for optical quantum communications, optical quantum sensors and even optical quantum simulators. On the other hand, IExG is a passive platform, that is, it has not electro-optical properties and presents a low nonlinearity, although efforts based on glass poling [188] and/or hybrid integration [189] have been made to achieve significant nonlinear effects. Therefore, in a monolithic version, IExG platform would perform only passive functions, but, as shown in Section 3, doping with rare-earth ions adds active functionality [60] and opens new possibilities for integrated quantum photonics. We must also underline that thermo-optic reconfiguration would also be possible in suitable glasses, so providing an additional capability. Accordingly, while applications to quantum computing, as for example the implementation of complex quantum processors for specific or universal purpose, are reserved for other platforms (as lithium niobate, silicon-on-insulator, ...), applications requiring a less amount of qubits, as quantum cryptography, quantum optical sensing and even quantum optical simulating, could be implemented by using the IExG platform. A few examples have already appeared in the scientific literature, and the primary aim of this section is to justify such possible applications.

Indeed, recently, theoretical and experimental studies of the IExG platform for implementing different integrated quantum devices has begun to be reported. Thus, numerical modeling of ion exchange in glass for quantum computations tasks has been discussed [190,191]; nanoemitters (nanodiamond) coupled to ion-exchanged glass waveguides for quantum photonics have been realized [192,193]; ion-exchanged glass directional couplers for implementing quantum projectors have been proposed [194]. On the other hand, ion-exchanged waveguides in thin glass substrates, suitable for photonic system integration and EOCB (electro-optical circuit board) panel size integration [195], can open new possibilities in IQP. Likewise, other proposed quantum systems could be implemented, as, for example, quantum simulators of geometrical phases [72] and boson sampling [196], quantum random walk devices [197] and so on. As commented, these implementations will be efficient as long as a few qubits are considered, a limit which is in turn related to the present scalability difficulty of quantum integrated photonics. Thus, ion-exchanged glass devices can provide optimal solutions in quantum optical communications, in particular integrated quantum devices for quantum cryptography [49,51,198] as for example generators of quantum states, quantum projectors and Bell states measurement devices. Likewise, experimental results of these devices fabricated by ion exchange in glass provide proofs of concept about the use of ion-exchanged glass integrated optics in quantum information; but as it will be shown these devices would use well-known ion-exchanged glass integrated optical elements as directional couplers (see, for instance, References [199–201]), integrated optical gratings (see, for instance, references [17–19]) and so on. Therefore, we have many of the basic ingredients to develop IQP in a IExG platform and we only need to design such quantum devices (quantum circuits), fabricate them and optimize their performance. For example, preliminary results of a quantum projector fabricated by Na^+/K^+ ion exchange in soda-lime glass has been recently proposed [194].

In all cases, the quantum circuits consist mainly of N channel waveguides or paths (N-dimensional device) formed by a set of concatenated integrated optical elements as directional couplers and phase shifters [49] performing quantum operations. As mentioned passive integrated optical elements allow the implementation of devices to be mainly used in quantum communications through different kinds of optical fibers as multicore optical fibers [50] and few-mode optical fibers [52]. On the other hand, it may be useful to recall that space multiplexing is one of the most promising method for increasing the transmission capacity of optical fiber systems [202,203] and that spatial multiplexing can be used for implementing high-dimensional quantum cryptography. At this regard, ion-exchanged glass modal converters have been already successfully used for spatial multiplexing in few-mode optical fibers [52,204–209]. These components are located at the emitter and receiver, where one can already place the ion-exchanged glass IQP devices to generate or measure quantum light states excited in the optical modes of a few-mode optical fiber.

Regarding the manufacturing of these devices, we must stress, as commented, that generation and measurement of quantum states in quantum communications can be made with devices formed by the same integrated optical elements fabricated on IExG platform and used in classical optical communications (directional couplers, $1 \times N$ splitters, optical gratings and so on). Therefore, from an industrial point of view, the existing manufacturers of photonic devices on IExG platform can easily enlarge their production to devices designed for quantum optical communications, or even other applications as optical quantum sensors and optical quantum simulators. These applications are well considered by many companies, and governments all over the world invest in quantum technologies that have a disruptive potential. For example, the European Commission has launched in 2018 an ambitious flagship initiative for quantum technologies [210] where, for example, quantum communications play a central role [211,212] for both increasing commercialization of research results and the interest of large industrial players. On the market side, a recent report by Allied Market Research estimated that the market size of integrated quantum optical circuits could grow from a value of 426.0 million USD in 2017 up to 1460.2 million USD by 2025 [213].

In order to make clear the above considerations, and after describing, for the sake of completeness, the fundamental aspects of quantum integrated optical elements in Section 4.1, we present, in Section 4.2, an integrated generator of quantum light states $|L\rangle$ based on the use of integrated optical lasers and directional couplers and mainly intended for quantum cryptography. Such a quantum states generator device is designed for the BB84 quantum key distribution scheme [214] and with two-dimensional quantum states, although it can be extended to N-dimensional quantum states [51]. In this device, we propose to use a rare-earth-doped glass, as discussed in Section 3, for generating four weak laser signals obtained by pumping beams, that is, to use a hybrid glass. On the other hand, the passive part of the above device can in turn be used as an integrated quantum projector, of a high interest in quantum devices, particularly and once again, in quantum cryptography [215,216]. Moreover, another device for measuring Bell states is also described, which is of a great importance in MDI (measurement device independent) protocol [217], which is based on the use of two-photon states unlike the BB84 protocol that uses a single photon. Experimental results on a single photon two-dimensional quantum projector is presented in Section 4.3.

4.1. Fundamental Aspects of Quantum Integrated Optical Components

The use of quantum light states in integrated optical devices allows to develop different on-chip applications in quantum optical communications and in other fields such as quantum photonic simulations [196] and quantum photonic sensing [218]. Rigorous theories of linear and nonlinear quantum spatial propagation states in integrated waveguides can be used [219–222], which allows us to study and design linear and nonlinear quantum integrated optical elements. Moreover, we consider three types of quantum states of a great interest for proofs of concept as the coherent states, the single photon states and the biphoton states. Single photon states, and in particular 1-qubits and 1-ququart states (dimension $d = 4$), have a high interest, for instance, in quantum cryptography; 1-ququart are the states giving rise to the beginning of the so called high dimension quantum cryptography with 1-qudits ($d = N > 2$), that is,

$$|L_s\rangle = \sum_{j=1}^{N} c_j|0\ldots 1_j\ldots 0\rangle \equiv \sum_{j=1}^{N} c_j|1_j\rangle \tag{4}$$

where subscript s stands for single-mode photon. This single photon state $|L_s\rangle$ shows that the photon is in a quantum superposition in the N channel waveguides. Moreover, by using 1-qudits states we can form quantum bases of orthogonal states. These bases have to be in turn mutually unbiased bases (MUBs), which requires $|c_1| = \ldots = |c_N|$ [198]. On the other hand, lasers emit coherent states $|\alpha\rangle$, with $\alpha \in \mathbb{C}$, then by using, for instance, a 1xN splitter or a cascade of directional couplers [51] a multimode coherent state $|\alpha_1 \ldots \alpha_N\rangle$ is obtained. It is well known, however, that for weak coherent (wc) states regime, $|\alpha_i| \ll 1$, an approximated multimode single photon state is achieved, that is, $|L_{wc}\rangle \approx |0\ldots 0\rangle + \sum_{j=1}^{N} \alpha_j|0\ldots 1_j\ldots 0\rangle$. These states are less efficient because there is a single photon state with a small probability, but it is enough for quantum cryptography.

Next, it is worth describing the quantum mechanical treatment of the basic passive integrated optical elements of our integrated circuits, that is, the directional couplers 2×2 and phase shifters. We must stress that these 2×2 integrated optical elements (couplers and phase shifters) are enough to generate unitary transformations of dimension N, that is, $SU(N)$ transformations, by concatening properly such optical elements [216]. Let us start by considering a two-dimensional quantum problem ($N = 2$), that is, the single photon quantum states are 1-qubits, namely, $|L_s\rangle = c_{o1}|1_1\rangle + c_{o2}|1_2\rangle$, with 1 and 2 denoting two single mode channel waveguides. Synchronous directional couplers can be represented by the following matrix

$$X_\theta = \begin{pmatrix} \cos\theta & i\sin\theta \\ i\sin\theta & \cos\theta \end{pmatrix} \tag{5}$$

where $\theta = \kappa L$, with κ the coupling coefficient and L the coupling (real or effective) length of the coupler. For single photon states, the standard algebra for a two-dimensional complex space can be applied, therefore the output state (or output 1-qubit) is given by the expression

$$|L'\rangle = (c_{o1}\cos\theta + ic_{o2}\sin\theta)|1_1\rangle + (ic_{o1}\sin\theta + c_{o2}\cos\theta)|1_2\rangle \equiv c_1|1_1\rangle + c_2|1_2\rangle \tag{6}$$

For quantum information applications the usually required couplers are $X_{\pi/4}$, that is, 3 dB (or 1:1) couplers, which correlate to the logic gate $H_c = \sqrt{-iX}$, (circular Hadamard logic gate) and $X_{\pi/2}$ corresponding to the usual Pauli's logic gate X. On the other hand, phase shifters are also required. Their implementation can be obtained by using different strategies (change in refractive index, change in length or width of a single-mode channel guide and so on) and they can even be reconfigurable by using glass thermo-optic modulation. In any case, the matrix representation of a phase shifter is given by

$$Z_\Phi = \begin{pmatrix} e^{-i\Phi/2} & 0 \\ 0 & e^{i\Phi/2} \end{pmatrix} \tag{7}$$

Phase shifters such as $Z_{\pi/2}$ and Z_π are usually required. In short, with several optical integrated elements X_θ and Z_Φ concatenated, along with active components, it is possible to implement different integrated optical circuits for use in quantum (and classical) optical communications with arbitary dimension N.

Finally, there are biphoton states, which can be obtained from an SPDC (spontaneous parametric down-conversion) quantum source. This source can emit twin photons or entangled photons. In a compact form, we can write a biphoton state as follows

$$|L_b\rangle = c_{jj'}|1_j 1_{j'}\rangle + c_{ll'}|1_l 1_{l'}\rangle \tag{8}$$

where j, l, j', l' indicate four different single-mode channel guides. If $c_{ll'} = 0$ then $c_{jj'} = 1$ and twin photons states are obtained. Likewise, if $c_{jj'} = \pm c_{ll'} = 1/\sqrt{2}$, the Bell states $|\Phi^\pm\rangle$ are obtained. If we perform the change $j' \leftrightarrow l'$ and $c_{jl'} = \pm c_{lj'} = 1/\sqrt{2}$ then the Bell states $|\Psi^\pm\rangle$ are obtained. Note that we obtain quantum states by using excitations in the modes of several single-mode channel waveguides, which can also be denominated as path modes or codirectional modes.

4.2. Integrated Quantum Circuits for Quantum Cryptography

In this subsection we present two examples of integrated quantum circuits. The first one is a generator of quantum states intended for quantum cryptography by using the so-called BB84 protocol [214]. It is made on a hybrid glass as the ones described in Sections 2 and 3. Moroever, we show that the passive part of this integrated optical circuit can be also used as a quantum projector. The circuit is made with four paths but can be generalized to an arbitrary number of channel guides. The second integrated circuit is a Bell states measurement device of interest for the so-called MDI (measurement device independent) protocol [217] or in general for measuring the Bell states $\Psi^\pm$.

4.2.1. Quantum States Generator and Quantum Projectors On-Chip for BB84

Figure 13 shows a device to generate four quantum states, which are required in the Alice system of the BB84 protocol. It consists of four paths, that is, four ion-exchanged channel guides fabricated in a hybrid glass substrate. Moreover, an OG in the rare-earth-doped glass section (DBR lasers) generates four weak laser signals (weak coherent states) obtained with pumping beams P_j, ($j = 1, 2, 3, 4$), whereas in the passive glass section four directional couplers $X_{\pi/4}$ and one coupler $X_{\pi/2}$ along with a phase shifter $Z_{\pi/2}$ generate the four states required for BB84 protocol at the outputs 2 and 3, namely, the 1-qubits $(1/\sqrt{2})(|1_2\rangle + e^{i\delta}|1_3\rangle)$, with $\delta = \pm\pi/2, \pi, 0$. The pumping beams P_j are activated in a random way, and moreover, as shown later, the device (Alice device) emits each of the four states through the outputs of paths 2 and 3 with a probability equal to 50%.

For example, if a weak coherent state is excited in channel waveguide 1 with pumping beam P_1, then it is obtained $|L\rangle \approx |1_1\rangle$. Next, the concatenated couplers $X_{\pi/4}$ and $X_{\pi/2}$ along with the shifter $Z_{\pi/2}$ (see Figure 13) perform a linear transformation, therefore, by taking into account Equation (5) for $\theta = \pi/4$ and $\theta = \pi/2$ and Equation (7) for $\Phi = \pi/2$, the following output state is obtained

$$|L_1\rangle = \frac{1}{2}(|1_1\rangle - i|1_4\rangle) + \frac{i}{2}(|1_2\rangle + i|1_3\rangle) \tag{9}$$

Note that in channel guides 2 and 3 we have one of the states of the BB84 protocol, that is, in a vector form we have the state $(1/\sqrt{2})(1, i)^t$, with t indicating transposed. The other three states are obtained by using the other pumping beams. All of them are produced with a 50% of probability what is reasonable for quantum key distributions.

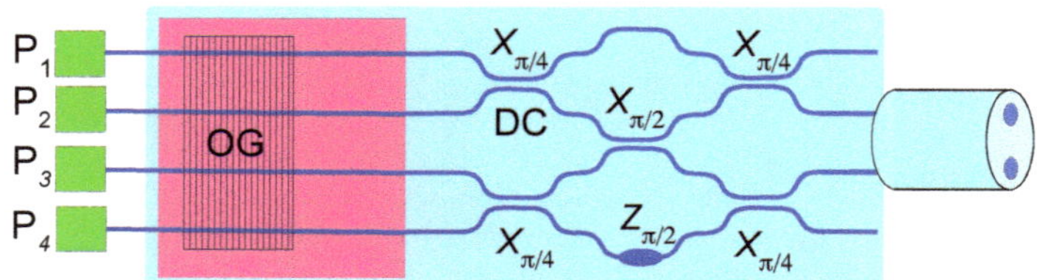

Figure 13. An integrated device with 4 paths (single-mode channel guides) on a hybrid glass substrate is shown. An optical grating (OG) in the active glass section generates four laser weak signals obtained with pumping beams P_j, $(j = 1, 2, 3, 4)$; in the passive glass section, four directional couplers $X_{\pi/4}$ and one coupler $X_{\pi/2}$ along with a phase shifter $Z_{\pi/2}$ generate the four states required for BB84 protocol at the outputs 2 and 3 coupled, for example, to two single-mode cores of a MCF.

On the other hand, the passive part of this device (the blue section, that is, without the active section where the OG is placed) is a quantum projector for the above states. Indeed, by considering that the inputs are the channel guides 1 and 3, then the first two couplers $X_{\pi/4}$ and the coupler $X_{\pi/2}$ make a state division. The next upper coupler $X_{\pi/4}$ projects the states $(1/\sqrt{2})(1, \pm 1)^t$, whereas the phase shifter and the bottom coupler $X_{\pi/4}Z_{\pi/2}$ project the states $(1/\sqrt{2})(1, \pm i)^t$. This means that the device (Bob device) produces a random choice of bases. For example, let us consider the state $(1/\sqrt{2})(|1_1\rangle + |1_3\rangle) \equiv (1/\sqrt{2})(1, 1)^t$, then, the final state is

$$|L_1\rangle = \frac{1}{\sqrt{2}}|1_2\rangle + \frac{1}{2}e^{-i\pi/4}|1_3\rangle + \frac{1}{2}e^{i\pi/4}|1_4\rangle \tag{10}$$

Note that if the state chooses the channel 1 and 2, then it is projected on the output 2 ($|1_2\rangle$). If the input state is $(1/\sqrt{2})(|1_1\rangle - |1_3\rangle) \equiv (1/\sqrt{2})(1, -1)^t$, then we obtain

$$|L_1\rangle = \frac{i}{\sqrt{2}}|1_1\rangle - \frac{1}{2}e^{-i\pi/4}|1_3\rangle + \frac{1}{2}e^{i\pi/4}|1_4\rangle \tag{11}$$

namely, the state is projected on the output 1 ($|1_1\rangle$). The same procedure can be followed for the states $(1/\sqrt{2})(|1_1\rangle \pm i|1_3\rangle) \equiv (1/\sqrt{2})(1, \pm i)^t$ which are projected on the outputs 3 and 4.

4.2.2. Bell States Measurement Integrated Device for MDI

Another interesting quantum passive device is a Bell states measurement device, which can be required in many optical quantum systems, e.g., in a MDI quantum cryptography system [217] in multicore optical fibers or few-mode optical fibers. This device can be implemented with integrated channel guides as shown next.

In Figure 14 a sketch of this device is given for the case of two multicore optical fibers F_a and F_b, where only two cores are shown. We must recall that multicore optical fibers and few-mode optical fibers allow for the increase of the optical transmission capability, therefore the use of these optical fibers for quantum communications is highly likely. Accordingly, the study of quantum operations with this kind of fiber also has a major potential interest. The Bell projector can be implemented by a $X_{\pi/2}$ coupler and two $X_{\pi/4}$ couplers. Indeed, let us consider the following biphoton (or 2-qubit) Bell states, according to the notation used in Figure 14,

$$\Psi^{\pm} = \frac{1}{\sqrt{2}}(|1_{ao1}1_{bo2}\rangle \pm |1_{ao2}1_{bo1}\rangle) \tag{12}$$

By applying the optical transformations produced by $X_{\pi/2}$ and $X_{\pi/4}$ couplers for single-photon states given by Equations (5) and (6) one obtains

$$\Psi^{+} \rightarrow \frac{i}{\sqrt{2}}(|1_{a1}1_{b1}\rangle + |1_{a2}1_{b2}\rangle), \quad \Psi^{-} \rightarrow \frac{1}{\sqrt{2}}(|1_{a1}1_{b2}\rangle + |1_{a2}1_{b1}\rangle) \tag{13}$$

Thus, when photons are registered in detectors D_{a1}, D_{b1} or D_{a2}, D_{b2} (photon coincidences) we have the Bell state Ψ^{+}, and when photons are registered in detectors D_{a1}, D_{b2} or D_{a2}, D_{b1} we have the Bell state Ψ^{-}. In short, an integrated Bell projector device has been obtained for Bell states $\Psi^{\pm}$.

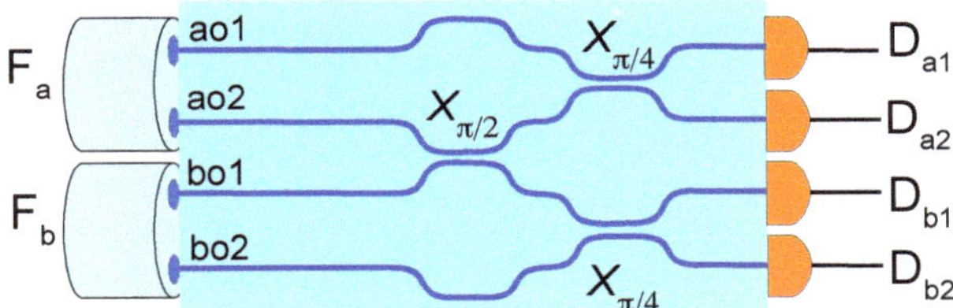

Figure 14. Integrated Bell projector for entangled quantum states excited in four cores $ao\,j$, $j = 1,2,3,4$ of two-core optical fibers F_a and F_b. Detectors D_{aj} are used for measuring photon coincidences.

On the other hand, this result has also an important impact in cryptography, in particular in a MDI quantum cryptography system. Indeed, let us consider the biphoton state $|1_{ao1}1_{bo2}\rangle$ launched by Alice and Bob, then the output state is given by

$$|L\rangle = \frac{1}{2}\{(|1_{a1}1_{b2}\rangle - (|1_{a2}1_{b1}\rangle)) + i(|1_{a1}1_{b1}\rangle + |1_{a2}1_{b2}\rangle)\} \equiv \frac{1}{\sqrt{2}}(\Psi^{-} + i\Phi^{+}) \tag{14}$$

that is, Bell states are again detected by photon concidences (Charlie device). Now, if the state launched by Alice and Bob is $|1_{ao2}1_{bo1}\rangle$, the result is the same, except the sign: $(1/\sqrt{2})(-\Psi^{-} + i\Phi^{+})$; therefore, a quantum key distribution is possible because Eve is not able to know which state has been sent, that is, $|1_{ao1}1_{bo2}\rangle$ or $|1_{ao2}1_{bo1}\rangle$.

The above examples (quantum states generator and quantum projectors for BB84 and a Bell states measurement device) have made clear the potential of the integrated optical devices for the implementation of quantum operations. Next, we show how the practical implementation of these devices in a ion-exchanged glass platform is possible because only a few qubits (or qudits) are required.

4.3. Fabrication and Characterization of a Quantum Projector

We present the fabrication of the basic units $X_{\pi/4}$ and $X_{\pi/2}$, that is, 2×2 directional couplers in glass, by using two consecutive purely thermal K^{+}/Na^{+} ion-exchange processes in glass. The first process is a selective ion exchange, that is, it is produced only in non-masked regions that corresponds to the channel waveguides defining the integrated optical component, while the second one is only a burial process through the same mask [194]. We must underline that this two-step process is different from the one proposed in Reference [199] where two consecutive K^{+}/Na^{+} ion exchanges are made, but the second is a planar ion exchange. Rigorously, the second step (burial process) would not be necessary but presents two important advantages. On the one hand, the modal field is located further from the glass surface, which can present irregularities, and therefore surface losses are reduced, and, on the other hand, it decreases the typical anisotropy in K^+/Na^+ waveguides produced by mechanical stress due to the size difference of the two ions, one of the main drawback of these integrated guides. Of course there are many other strategies for fabricating directional couplers by K^+/Na^+ ion exchange (see, for example [16,200,201]), which could also be suitable for this kind of applications. Another approach would be to use ion-exchanged MMI couplers based on multimode interference [223,224], which would have less stringent lithographic requirements. There are also proposals of reconfigurable integrated elements as directional couplers by thermo-optic modulation [225], a strategy that can be oriented to reconfigure quantum operations. In another work, an interesting design of asymmetric Y-branch waveguides for bidirectional transmission on PCB (printed circuit board) was described [226], which would allow us to join IQP and microelectronics.

The fabrication strategy presented here has given excellent experimental results for $X_{\pi/4}$ and $X_{\pi/2}$ directional couplers with both low losses and anisotropic effects (coupling between TE and TM guided modes, one of the drawbacks of K^+/Na^+ ion exchange) thanks to the burial process. The starting point has been to use previous results [54,64,201,207,227], which have allowed us to establish a fixed temperature $T = 400\,°C$ for both the first and the second (burial) ion exchange in a soda-lime glass with index $n_s = 1.5104$ at $\lambda_o = 633\,nm$. Several tests of planar thermal ion exchange in the same type of glass substrate were made, and the effective indices of the different guided modes were measured by prism coupling method in an automated system (Metricon, Model 2010/M). From this empiric calibration and taking into account numerical results by using the EIM for channel guides, a diffusion time for the first exchange equal to $t_1 = 30\,min$ and a time $t_2 = 10\,min$ for the burial process were fixed in order to achieve single-mode channel waveguides.

In order to design and fabricate integrated quantum optical projectors, a prior optical characterization of the full fabrication process of the basic elements (the directional couplers as $X_{\pi/4}$ and $X_{\pi/2}$) was needed. A series of masks on the same substrate for different directional couplers have been made by changing the separation s between channel guides and the coupling length l of the couplers (a sketch of the directional coupler is shown in Figure 15), while keeping the width of the channel guides constant ($w = 3\,\mu m$). The connection between channel guides have been made through curves following the profile used by Minford et al. (MC curve) [62]. The separations and the coupling lengths chosen, after the photo-reduction process, are found in the interval $s \in [3,6]\,\mu m$ and $l \in [0.5,2]\,mm$, respectively. We must indicate that under the ion-exchange process, the lateral diffusion increases the width of the channel guides up to $4\,\mu m$ with respect to the designed mask width of $3\,\mu m$.

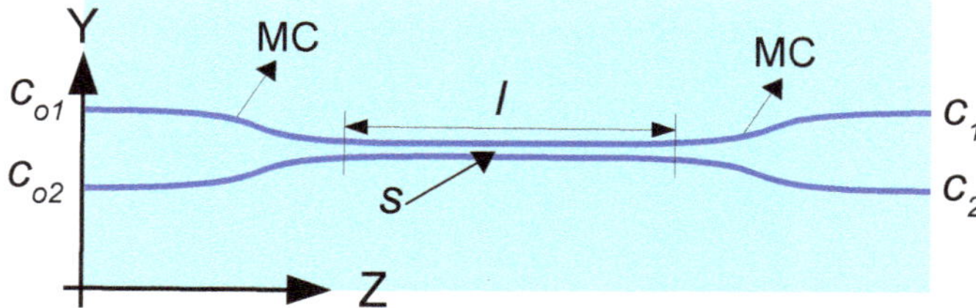

Figure 15. Sketch of a directional coupler belonging to the series of directional couplers fabricated. The main parameters are indicated: separation s between channel guides and length l of the guides from after and up to before the Minford curves (MC).

An optical characterization of the directional couplers was then made. We denote as c_{o1} and c_{o2}, respectively, the input complex amplitudes at the two guides of the coupler and as c_1 and c_2 the corresponding output complex amplitudes, so that the output powers are $P_i = |c_i|^2$, $i = 1, 2$. In the tests, light was coupled to one of the branches of the coupler and the power at each output was measured. Experimental results are shown in Table 1.

Some important conclusions can be obtained from the results shown in Table 1: the coupler $(s, l) = (3.0, 2.0)$ is very close to a $X_{\pi/2}$ coupler, and the coupler $(s, l) = (6.0, 1.5)$ is very close to a $X_{\pi/4}$ coupler.

Table 1. Normalized powers $P_1|P_2$ at coupler's outputs 1 and 2, respectively.

Series of Couplers		l (mm)			
		0.5	1.0	1.5	2.0
s (μm)	3.0	(17.8|82.2)	(93.6| 6.4)	(54.9|45.1)	(2.5|97.5)
	4.5	(35.9|64.1)	(5.9|94.1)	(7.4|92.6)	(40.9|59.1)
	6.0	(80.7|19.3)	(63.8|36.2)	(49.3|50.7)	(26.8|73.2)

Alternatively, one can use a numerical linear fitting to obtain the optimal values of (s, l) for fabricating a particular coupler X_θ; thus, Figure 16 shows the linear fit $\theta(l) = a\,l + b$ for the three separations s considered. The term $\theta(0)$ is the coupling introduced by the Minford curves and it is different from zero because l is the distance between the curved sections. Note that the above fit allows us to define an effective length l_{eff} such that $\theta(l_{\text{eff}}) = \kappa l_{\text{eff}}$, where κ is the coupling coefficient ($a \equiv \kappa$), and thus we can write the following relationship

$$\theta(l) = \kappa l + \theta(0) \tag{15}$$

By deriving κ and $\theta(0)$ from the linear fit of the experimental coupling results we obtain the effective length $l_{\text{eff}} = l + \theta(0)/\kappa$. In short, we have found the empiric parameters for fabricating simultaneously $X_{\pi/2}$ and $X_{\pi/2}$ couplers and thus to fabricate the three IQP devices described in the above subsection. The phase shifter $\pi/2$ could be implemented by changing in an adiabatic way the length or the width of a channel guide.

The experimental results for the directional coupler $(s, l) = (6.0, 1.5)$ ($X_{\pi/4}$ coupler) as a quantum projector can now be shown. The test was made by using a He-Ne laser source, so that we made a semi-classical optical characterization with coherent states, which is sufficient to characterize the directional coupler; we must stress, however, that with single-photon states the same results would be obtained. We excited different states at the input of the directional coupler by using an optical grating [194]. Indeed, let us consider an external optical grating with a periodic structure along the y axis (see, for instance, Figure 15). When it is normally illuminated by a Gaussian beam propagating along z direction, and the diffracted light is collected by a lens, then the diffraction orders ± 1 can be coupled to the inputs c_{o1} and c_{o2} of the directional coupler shown in Figure 15. Next, by translating the optical grating along y axis, different relative phases δ between the amplitudes are obtained; in other words, the following input vector of amplitudes is obtained at the directional coupler: $v_o^t = (c_{o1}, e^{i\delta} c_{o2})^t$, where superscript t stands for transposed. It is worth writing the corresponding input quantum state, i.e., the bimode coherent state $|c_{o1} \ e^{i\delta} c_{o2}\rangle$. Next, we choose the directional coupler $(s, l) = (6.0, 1.5)$, which is very close to a $X_{\pi/4}$ coupler and therefore is a projector of quantum states $(1/\sqrt{2})(|1_1\rangle \pm i|1_2\rangle)$. Alternatively, the device formed by an optical grating introducing a $\pi/2$ phase plus a directional coupler $X_{\pi/4}$ can be formally regarded as a projector for states $(1/\sqrt{2})(|1_1\rangle \pm |1_2\rangle)$. Note that if we use an active optical grating implemented by a spatial light modulator introducing 0 or $\pi/2$ in a random way we have a quantum states detection system (Bob) of random bases.

In Figure 17, the experimental results (image of waveguide outputs) for these projections are shown, with the corresponding input coherent states. Note that the bimode coherent states $|\alpha/\sqrt{2} \pm i\alpha/\sqrt{2}\rangle$ are projected in states $|\alpha 0\rangle$ and $|0\alpha\rangle$, therefore single photon states $(1/\sqrt{2})(|1_1\rangle \pm i|1_2\rangle)$, will be projected in states $|1_1\rangle$ and $|1_2\rangle$, respectively. On the other hand, the following two-mode coherent states $|\alpha/\sqrt{2} \pm \alpha/\sqrt{2}\rangle$ do not undergo transformations, therefore the single photon states $(1/\sqrt{2})(|1_1\rangle \pm |1_2\rangle)$ also do not, and consequently it is also experimentally proved that they are eigenstates. This last case is highly interesting to implement a quantum random number generator.

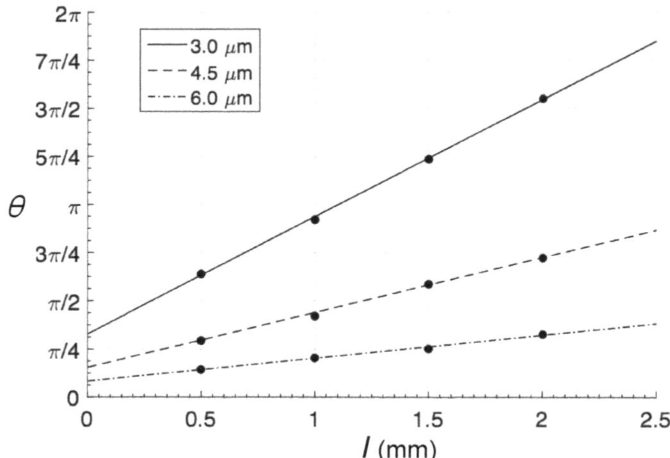

Figure 16. Linear fits of the coupling $\theta(l)$ for the three separations s between channel guides of twelve directional couplers.

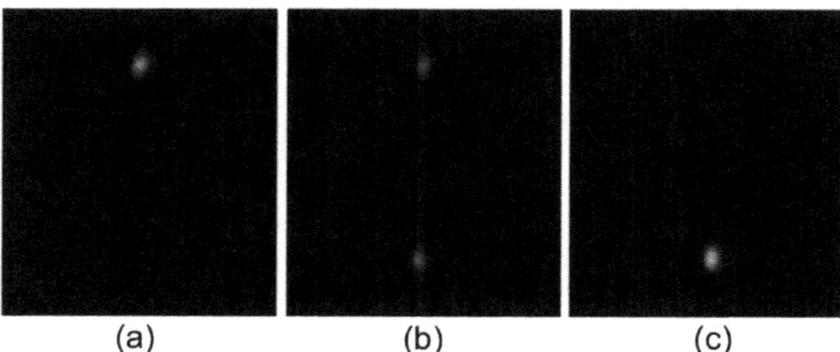

Figure 17. Experimental outputs for different input states to a coupler $X_{\pi/4}$. (a) Input state: $|\alpha/\sqrt{2} \ -i\alpha/\sqrt{2}\rangle$; output state: $|\alpha\ 0\rangle$; (b) input and output states: $|\alpha/\sqrt{2}\ \pm\alpha/\sqrt{2}\rangle$; (c) input state: $|\alpha/\sqrt{2}\ i\alpha/\sqrt{2}\rangle$; output state: $|0\ \alpha\rangle$.

In short, the experimental results achieved with the directional coupler confirm that a $X_{\pi/4}$ transformation is implemented, and therefore a projection operation can be made. A similar experimental study has been made with the $X_{\pi/2}$ coupler. Therefore, when several couplers $X_{\pi/4}$ and $X_{\pi/2}$ are combined, that is, couplers $(s,l) = \{(6.0, 1.5), (3.0, 2.0)\}$ fabricated simultaneously, the devices described in Section 4.2 can be fabricated and also more general projectors (high-dimensional projectors, Bell states projectors, . . .) by using ion exchange in glass.

5. Conclusions

In this work we have presented a review on active and quantum integrated photonic elements fabricated by ion exchange in glass, which have a remarkable impact in different fields such as classical and quantum optical communications, optical processing and optical sensing. Ion-exchanged components in rare-earth-doped glasses provide an optimal solution to implement integrated optical lasers and amplifiers; erbium-doped waveguide amplifiers represent a very important element in optical transmission systems. Likewise, a growing interest has emerged for the use of the ion-exchanged waveguide technology as an effective platform for integrated quantum photonics. We have reported theoretical and experimental results showing the high potentiality of this platform for developing active and integrated quantum devices, with particular reference to the combination of active and passive ion-exchanged glass elements in hybrid glasses. Thus, we have shown how to implement, for example, an active integrated quantum states generator device for quantum cryptography (Alice device) and passive devices such as, for example, an integrated quantum projector, suitable also for the BB84 protocol in quantum cryptography (Bob device), which allows us to detect quantum states by projective measurements, and a Bell states measurement device for MDI protocol in quantum cryptography (Charlie device). As an overall, we have shown that ion exchange in glass, even if being a very ancient technology, is still crucial for today's (and future) photonic and quantum integration technology.

Author Contributions: The authors contributed equally to the manuscript. Both have read and agreed to the published version of the manuscript.

Funding: This research was funded by Xunta de Galicia, Consellería de Educación, Universidades e FP, Grant GRC Number ED431C2018/11.

Data Availability Statement: Data is contained within the article.

Conflicts of Interest: The authors declare no conflict of interest.

Abbreviations

The following abbreviations are used in this manuscript:

AS	Aluminosilicate
ASE	Amplified Spontaneous Emission
BB84	Bennett and Brassard 1984
CMOS	Complementary Meta Oxide Semiconductor
DC	Directional Coupler
EDFA	Er-Doped Fiber Amplifier
EDWA	Erbium-Doped Waveguide Amplifier
EYDWA	Erbium-Ytterbium-Doped Waveguide Amplifier
EOCB	Electro-Optical Circuit Board
FMF	Few-Mode Optical Fiber
IExG	Ion-Exchanged Glass
IQP	Integrated Quantum Photonics
MC	Minford et al. Curve
MCF	Multicore Optical Fiber
MDI	Measurement Device Independent
MMI	Multi-Mode Interference
MUBs	Mutually Unbiased Bases
NLO	Nonlinear optics
OG	Optical Grating
PCB	Printed Circuit Board
PECVD	Plasma-Enhanced Chemical Vapor Deposition
RED	Rare-Earth Doped
SPDC	Spontaneous Parametric Down-Conversion)
SL	Soda–Lima–silicate
WDM	Wavelength Division Multiplexing

References

1. Pezzella, S. *Il Trattato di Antonio da Pisa Sulla Fabbricazione Delle Vetrate Artistiche*; Umbria Edizioni: Perugia, Italy, 1976.
2. Schulze, G. Versuche über die diffusion von silber in glas. *Angew. Physik* **1913**, *40*, 335–367. [CrossRef]
3. Righini, G.C.; Pelli, S. *1969–2019: 50 Years of Integrated Optics*; Integrated Optics; Righini, G.C., Ferrari, M., Eds.; The IET: London, UK, 2020; Volume 1, pp. 1–38.
4. Izawa, T.; Nakagome, H. Optical waveguide formed by electrically induced migration of ions in glass plates. *Appl. Phys. Lett.* **1972**, *21*, 584–586. [CrossRef]
5. Giallorenzi, T.G.; West, E.J.; Kirk, R.; Ginther, R.; Andrews, R.A. Optical waveguides formed by thermal migration of ions in glass. *Appl. Opt.* **1973**, *12*, 1240–1245. [CrossRef]

6. Koizumi, K.; Sumimoto, T.; Matsushita, S.; Furukawa, M. Ion Exchange Process for Manufacturing Integrated Optical Circuits. U.S. Patent N. 3,857,689, 31 December 1974.

7. Sumimoto, T.; Matsushita, S.; Yamazaki, T.; Fujiwara, T.; Koizumi, K. Method for Manufacturing Optical Integrated Circuits Utilizing an External Electric Field. U.S. Patent N. 3,836,348A, 17 September 1974.

8. Najafi, S.I. (Ed.) *Glass Integrated Optics and Optical Fiber Devices*; Number CR53 in Critical Reviews of Optical Science and Technology; SPIE: Bellingham, WA, USA, 1994.

9. Kita, H.; Kitano, I.; Uchida, T.; Furukawa, M. Light-Focusing Glass Fiber and Rods. *J. Am. Ceram. Soc.* **1971**, *54*, 321. [CrossRef]

10. Ohmi, S.; Sakai, H.; Asahara, Y.; Nakayama, S.; Yoneda, Y.; Izumitani, T. Gradient-index rod lens made by a double ion-exchange process. *Appl. Opt.* **1988**, *27*, 496–499. [CrossRef]

11. Liñares, J.; Montero, C.; Prieto, X. Graded-index bifocal spectacle lenses produced by ion-exchange in glass: Paraxial designing. *Pure Appl. Opt.* **1995**, *4*, 695–699. [CrossRef]

12. Gomez-Reino, C.; Perez, M.; Bao, C.; Flores-Arias, M. Design of GRIN optical components for coupling and interconnects. *Laser Photonics Rev.* **2008**, *2*, 203–215. [CrossRef]

13. Graydon, O. Graded lens surprise. *Nat. Photonics* **2019**, *13*, 732. [CrossRef]

14. Findakly, T. Glass waveguides by ion-exchange: A review. *Opt. Engin.* **1985**, *24*, 244–250. [CrossRef]

15. Ramaswamy, R.V.; Srivastava, R. Ion-exchanged glass waveguides: A review. *J. Light. Technol.* **1988**, *6*, 984–1000. [CrossRef]

16. Najafi, S.I. *Introduction to Glass Integrated Optics*; Artech House: Boston, MA, USA, 1986.

17. Righini, G.C.; Pelli, S. Ion exchange in glass: A mature technology for photonic devices. In *Materials and Devices for Photonic Circuits II*; Armenise, M.N., Ed.; SPIE: Bellingham, WA, USA, 2001; Volume 4453, pp. 93–99. [CrossRef]

18. Honkanen, S.; West, B.R.; Yliniemi, S.; Madasamy, P.; Morrell, M.; Auxier, J.; Schulzgen, A.; Peyghambarian, N.; Carriere, J.; Frantz, J.; et al. Recent advances in ion exchanged glass waveguides and devices. *Phys. Chem. Glas.* **2006**, *47*, 110–120.

19. Tervonen, A.; Honkanen, S.K.; West, B.R. Ion-exchanged glass waveguide technology: A review. *Opt. Eng.* **2011**, *50*, 1–16. [CrossRef]

20. Righini, G.C.; Chiappini, A. Glass optical waveguides: A review of fabrication techniques. *Opt. Eng.* **2014**, *53*, 071819. [CrossRef]

21. Fernandez, T.; Sakakura, M.; Eaton, S.; Sotillo, B.; Siegel, J.; Solis, J.; Shimotsuma, Y.; Miura, K. Bespoke photonic devices using ultrafast laser driven ion migration in glasses. *Prog. Mater. Sci.* **2018**, *94*, 68–113. [CrossRef]

22. Grelin, J.; Ghibaudo, E.; Broquin, J.E. Study of deeply buried waveguides: A way towards 3D integration. *Mater. Sci. Eng. B* **2008**, *149*, 185–189. [CrossRef]

23. Pei, C.; Wang, G.; Yang, B.; Yang, L.; Hao, Y.; Jiang, X.; Yang, J. Study of selectively buried ion-exchange glass waveguides using backside masking. *Chin. Opt. Lett.* **2015**, *13*, 021301.

24. Zhang, L.; Guo, X. Thermal history and its implications: A case study for ion exchange. *J. Am. Ceram. Soc.* **2020**, *103*, 3971–3977. [CrossRef]

25. Jordan, E.; Geoffray, F.; Bouchard, A.; Ghibaudo, E.; Broquin, J.E. Development of Tl+/Na+ ion-exchanged single-mode waveguides on silicate glass for visible-blue wavelengths applications. *Ceram. Int.* **2015**, *41*, 7996–8001. [CrossRef]

26. Liñares, J.; Montero, C.; Sotelo, D. Theory and design of an integrated optical sensor based on planar waveguiding lenses. *Opt. Commun.* **2000**, *180*, 29–36. [CrossRef]

27. Barral, D.; Liñares-Beiras, J.; Montero-Orille, C.; Moreno, V.; Prieto-Blanco, X.; Grazu, V.; Sánchez, C.; Seara, R. Integrated Ion-Exchanged Glass Bimodal Optical Biosensor based on Modal Interference Imaging. In Proceedings of the IX Optoelectronics Spanish Meeting OPTOEL 2015, Salamanca, Spain, 13–15 July 2015; p. PO–SII–35, ISBN 978-84-606-9716-9.

28. Asquini, R.; Buzzin, A.; Caputo, D.; de Cesare, G. Integrated Evanescent Waveguide Detector for Optical Sensing. *IEEE Trans. CPMT* **2018**, *8*, 1180–1186. [CrossRef]

29. Du, B.; Tong, Z.; Mu, X.; Xu, J.; Liu, S.; Liu, Z.; Cao, W.; Qi, Z.M. A Potassium Ion-Exchanged Glass Optical Waveguide Sensor Locally Coated with a Crystal Violet-SiO_2 Gel Film for Real-Time Detection of Organophosphorus Pesticides Simulant. *Sensors* **2019**, *19*, 4219-1–12. [CrossRef] [PubMed]

30. Mamtmin, G.; Abdurahman, R.; Yan, Y.; Nizamidin, P.; Yimit, A. A highly sensitive and selective optical waveguide sensor based on a porphyrin-coated ZnO film. *Sen. Actuators A* **2020**, *309*, 111918. [CrossRef]

31. Liu, L.; Zhou, X.; Wilkinson, J.S.; Hua, P.; Song, B.; Shi, H. Integrated optical waveguide-based fluorescent immunosensor for fast and sensitive detection of microcystin-LR in lakes: Optimization and Analysis. *Sci. Rep.* **2017**, *7*, 3655. [CrossRef] [PubMed]

32. Nikonorov, N.V.; Petrovskii, G.T. Ion-exchanged glasses in integrated optics: The current state of research and prospects (a review). *Glass Phys. Chem.* **1999**, *25*, 16–55.

33. Opilski, A.; Rogozinski, R.; Gut, K.; Blahut, M. Aspects of application of ion exchange in glass. In *Lightguides and Their Applications*; Wojcik, J., Wojcik, W., Eds.; SPIE: Bellingham, WA, USA, 2000; Volume 4239, pp. 1–10. [CrossRef]

34. Jackel, J.L.; Vogel, E.M.; Aitchison, J.S. Ion-exchanged optical waveguides for all-optical switching. *Appl. Opt.* **1990**, *29*, 3126–3129. [CrossRef]

35. Roussignol, P.; Ricard, D.; Flytzanis, C. Nonlinear optical properties of commercial semiconductor doped glasses. *Appl. Phys. A* **1987**, *44*, 285–292. [CrossRef]

36. Lawandy, N.M.; MacDonald, R.L. Optically encoded phase-matched second-harmonic generation in semiconductor-microcrystallite-doped glasses. *JOSA B* **1991**, *8*, 1307–1314. [CrossRef]

37. Righini, G.; Pelli, S.; Blasi, C.D.; Fagherazzi, G.; Manno, D. Ion-exchanged waveguides in semiconductor-doped glasses. In *Glasses for Optoelectronics II*; SPIE: Bellingham, WA, USA, 1991; Volume 1513, pp. 105–111. [CrossRef]
38. Charlet, B.; Bastard, L.; Broquin, J.E. 1 kW peak power passively Q-switched Nd^{3+}-doped glass integrated waveguide laser. *Opt. Lett.* **2011**, *36*, 1987–1989. [CrossRef]
39. Geoffray, F.; Bastard, L.; Broquin, J.E. High confinement ion-exchanged waveguides for nonlinear applications. *Ceram. Int.* **2015**, *41*, 8034–8039. [CrossRef]
40. Legrand, L.; Bouchard, A.; Grosa, G.; Broquin, J.E. Hybrid integration of 300 nm-thick $LiNbO_3$ films on ion-exchanged glass waveguides for efficient nonlinear integrated devices. In *Integrated Optics: Devices, Materials, and Technologies XXII*; García-Blanco, S.M., Cheben, P., Eds.; SPIE: Bellingham, WA, USA, 2018; Volume 10535, pp. 44–50. [CrossRef]
41. Editorial. The rise of integrated quantum photonics. *Nat. Photonics* **2020**, *14*, 265. [CrossRef]
42. Politi, A.; Crya, M.J.; Rarity, J.G.; Yu, S.; O'Brien, J.L. Silica-on-silicon waveguide quantum circuits. *Science* **2008**, *320*, 646–649. [CrossRef]
43. Politi, A.; Matthews, J.C.F.; Thompson, M.G.; O'Brien, J.L. Integrated Quantum Photonics. *IEEE J. Sel. Top. Quantum Electron.* **2009**, *15*, 1673–1684. [CrossRef]
44. Bogdanov, S.; Shalaginov, M.Y.; Boltasseva, A.; Shalaev, V.M. Material platforms for integrated quantum photonics. *Opt. Mater. Express* **2017**, *7*, 111–132. [CrossRef]
45. Wang, J.; Sciarrino, F.; Laing, A.; Thompson, M. Integrated photonic quantum technologies. *Nat. Photonics* **2020**, *14*, 273–284. [CrossRef]
46. Smith, D.H.; Mennea, P.L.; Gates, J.C. Integrated quantum photonics. In *Integrated Optics*; Righini, G.C., Ferrari, M., Eds.; The Institution of Engineering and Technology: London, UK, 2021; Volume 2, pp. 337–359.
47. Elshaari, A.W.; Pernice, W.; Srinivasan, K.; Benson, O.; Zwiller, V. Hybrid integrated quantum photonic circuits. *Nat. Photonics* **2015**, *14*, 285–298. [CrossRef]
48. Meany, T.; Gräfe, M.; Heilmann, R.; Perez-Leija, A.; Gross, S.; Steel, M.J.; Withford, M.J.; Szameit, A. Laser written circuits for quantum photonics. *Laser Photonics Rev.* **2015**, *9*, 363–384. [CrossRef]
49. Liñares, J.; Nistal, M.C.; Barral, D.; Moreno, V.; Montero, C.; Prieto, X. Quantum Integrated Optics: Theory and Applications. *Opt. Pura Apl.* **2011**, *44*, 241–253.
50. Xia, C.; Amezcua-Correa, R.; Bai, N.; Antonio-Lopez, E.; May-Arrioja, D.; Schulzgen, A.; Richardson, M.; Liñares, J.; Montero, C.; Mateo, E.; et al. Hole-Assisted Few-Mode Multicore Fiber for High-Density Space-Division Multiplexing. *IEEE Photonics Technol. Lett.* **2012**, *24*, 1914–1917. [CrossRef]
51. Balado, D.; Liñares, J.; Prieto-Blanco, X.; Barral, D. Phase and polarization autocompensating N-dimensional quantum cryptography in multicore optical fibers. *JOSA B* **2019**, *36*, 2793–2803. [CrossRef]
52. Bai, N.; Ip, E.; Huang, Y.K.; Mateo, E.; Yaman, F.; Li, M.J.; Bickham, S.; Ten, S.; Liñares, J.; Montero, C.; et al. Mode-division multiplexed transmission with inline few-mode fiber amplifier. *Opt. Express* **2012**, *20*, 2668–2680. [CrossRef]
53. Li, G.; Winick, K.A.; Griffin, H.C.; Hayden, J.S. Systematic modeling study of channel waveguide fabrication by thermal silver ion exchange. *Appl. Opt.* **2006**, *45*, 1743–1755. [CrossRef] [PubMed]
54. Prieto-Blanco, X.; Liñares, J. Two-Mode Waveguide Characterization by Intensity Measurements from Exit Face Images. *IEEE Photonics J.* **2012**, *4*, 65–79. [CrossRef]
55. Righini, G.C.; Pelli, S.; Ferrari, M.; Armellini, C.; Zampedri, L.; Tosello, C.; Ronchin, S.; Rolli, R.; Moser, E.; Montagna, M.; et al. Er-doped silica-based waveguides prepared by different techniques: RF-sputtering, sol-gel and ion-exchange. *Opt. Quantum Electron.* **2002**, *34*, 1151–1166. [CrossRef]
56. Luff, B.J.; Harris, R.D.; Wilkinson, J.S.; Wilson, R.; Schiffrin, D.J. Integrated-optical directional coupler biosensor. *Opt. Lett.* **1996**, *21*, 618–620. [CrossRef]
57. Tepper, J.; Labadie, L.; Diener, R.; Minardi, S.; Pott4, J.U.; Thomson, R.; Nolte, S. Integrated optics prototype beam combiner for long baseline interferometry in the L and M bands. *Astron. Astrophys.* **2017**, *602*, A66-1–A66-8. [CrossRef]
58. Conzone, S.D.; Hayden, J.S.; Funk, D.S.; Roshko, A.; Veasey, D.L. Hybrid glass substrates for waveguide device manufacture. *Opt. Lett.* **2001**, *26*, 509–511. [CrossRef]
59. Hayden, J.S.; Simpson, R.D.; Conzone, S.D.; Hickernell, R.K.; Callicoatt, B.; Roshko, A.; Sanford, N.A. Passive and active characterization of hybrid glass substrates for telecommunication applications. In *Rare-Earth-Doped Materials and Devices VI*; Jiang, S., Keys, R.W., Eds.; International Society for Optics and Photonics, SPIE: Bellingham, WA, USA, 2002; Volume 4645, pp. 43–50. [CrossRef]
60. Casale, M.; Bucci, D.; Bastard, L.; Broquin, J.E. Hybrid erbium-doped DFB waveguide laser made by wafer bonding of two ion-exchanged glasses. *Ceram. Int.* **2015**, *41*, 7466–7470. [CrossRef]
61. Blaize, S.; Bastard, L.; Cassagnetes, C.; Broquin, J.E. Multiwavelengths DFB waveguide laser arrays in Yb-Er codoped phosphate glass substrate. *IEEE Photonics Technol. Lett.* **2003**, *15*, 516–518. [CrossRef]
62. Minford, W.J.; Korotky, S.K.; Alferness, R.C. Low-Loss $Ti:LiNbO_3$ Waveguide Bends at λ_0 =1.3 µm. *IEEE J. Quantum Electron.* **1982**, *QE-18*, 1802–1906. [CrossRef]
63. Pelli, S.; Righini, G.C.; Scaglione, A.; Yip, G.L.; Noutsios, P.C.; Braeuer, A.H.; Dannberg, P.; Linares, J.; Reino, C.R.; Mazzi, G.; et al. Testing of optical waveguides (TOW) cooperative project: Preliminary results of the characterization of k-exchanged

waveguides. In *Linear and Nonlinear Integrated Optics*; Righini, G.C., Yevick, D., Eds.; SPIE: Bellingham, WA, USA, 1994; Volume 2212, pp. 126–131. [CrossRef]

64. Liñares, J.; Prieto, X.; Montero, C. A Novel Refractive Index Profile for Optical Characterization of Nonlinear Diffusion Processes and Planar Waveguides in Glass. *Opt. Mater.* **1994**, *3*, 229–236. [CrossRef]

65. Sebastiani, S.; Berneschi, S.; Brenci, M.; Nunzi-Conti, G.; Pelli, S.; Righini, G.C. Simple approach to calculate the refractive index profile of ion-exchanged waveguides. *Opt. Eng.* **2005**, *44*, 1–5. [CrossRef]

66. Liñares, J.; Moreno, V.; Nistal, M.C.; Salgueiro, J.R. Modelling of ion-exchanged monomode channel guides with quasi-exact modal solutions by the effective index method. *J. Mod. Opt.* **2001**, *48*, 789–795. [CrossRef]

67. Araujo, R. Interdiffusion in a one-dimensional interacting system. *J. Non-Cryst. Solids* **1993**, *152*, 70–74. [CrossRef]

68. Tervonen, A. A general model for fabrication processes of channel waveguides by ion exchange. *J. Appl. Phys.* **1990**, *67*, 2746–2752. [CrossRef]

69. Salmio, R.; Saarinen, J.; Turunen, J.; Tervonen, A. Graded-index diffractive elements by thermal ion exchange in glass. *Appl. Phys. Lett.* **1995**, *66*, 917–919. [CrossRef]

70. Prieto, X.; Liñares, J. Increasing resistivity effects in field-assisted ion exchange for planar optical waveguide fabrication. *Opt. Lett.* **1996**, *21*, 1363–1365. [CrossRef]

71. Prieto-Blanco, X. Electro-diffusion equations of monovalent cations in glass under charge neutrality approximation for optical waveguide fabrication. *Opt. Mater.* **2008**, *31*, 418–428. [CrossRef]

72. Liñares, J.; Prieto-Blanco, X.; Carral, G.M.; Nistal, M.C. Quantum Photonic Simulation of Spin-Magnetic Field Coupling and Atom-Optical Field Interaction. *Appl. Sci.* **2020**, *10*, 8850. [CrossRef]

73. Hanzawa, N.; Saitoh, K.; Sakamoto, T.; Matsui, T.; Tomita, S.; Koshiba, M. Demonstration of mode-division multiplexing transmission over 10 km two-mode fiber with mode coupler. In Proceedings of the 2011 Optical Fiber Communication Conference, Los Angeles, CA, USA, 6–10 March 2011; pp. 1–3.

74. Snitzer, E. Optical maser action of Nd^{3+} in a barium crown glass. *Phys. Rev. Lett.* **1961**, *7*, 444. [CrossRef]

75. Koester, C.J.; Snitzer, E. Amplification in a fiber laser. *Appl. Opt.* **1964**, *3*, 1182–1186. [CrossRef]

76. Yajima, H.; Kawase, S.; Sekimoto, Y. Amplification at 1.06 µm using a Nd: Glass thin-film waveguide. *Appl. Phys. Lett.* **1972**, *21*, 407–409. [CrossRef]

77. Saruwatari, M.; Izawa, T. Nd-glass laser with three-dimensional optical waveguide. *Appl. Phys. Lett.* **1974**, *24*, 603–605. [CrossRef]

78. Najafi, S.I.; Wang, W.; Currie, J.F.; Leonelli, R.; Brebner, J.L. Fabrication and characterization of neodymium-doped glass waveguides. *IEEE Photonics Technol. Lett.* **1989**, *1*, 109–110. [CrossRef]

79. Aoki, H.; Maruyama, O.; Asahara, Y. Glass waveguide laser. *IEEE Photonics Technol. Lett.* **1990**, *2*, 459–460. [CrossRef]

80. Aoki, H.; Ishikawa, E.; Asahara, Y. Nd^{3+}-doped glass waveguide amplifier at 1.054 µm. *Electron. Lett.* **1991**, *27*, 2351–2353. [CrossRef]

81. Mwarania, E.; Reekie, L.; Wang, J.; Wilkinson, J. Low-threshold monomode ion-exchanged waveguide lasers in neodymium-doped BK-7 glass. *Electron. Lett.* **1990**, *26*, 1317–1318. [CrossRef]

82. Sanford, N.A.; Malone, K.J.; Larson, D.R.; Hickernell, R.K. Y-branch waveguide glass laser and amplifier. *Opt. Lett.* **1991**, *16*, 1168–1170. [CrossRef] [PubMed]

83. Roman, J.E.; Winick, K.A. Neodymium-doped glass channel waveguide laser containing an integrated distributed Bragg reflector. *Appl. Phys. Lett.* **1992**, *61*, 2744–2746. [CrossRef]

84. Miliou, A.N.; Cao, X.F.; Srivastava, R.; Ramaswamy, R.V. 15-dB amplification at 1.06µm in ion-exchanged silicate glass waveguides. *IEEE Photonics Technol. Lett.* **1993**, *5*, 416–418. [CrossRef]

85. Aust, J.A.; Malone, K.J.; Veasey, D.L.; Sanford, N.A.; Roshko, A. Passively Q-switched Nd-doped waveguide laser. *Opt. Lett.* **1994**, *19*, 1849–1851. [CrossRef]

86. Mears, R.; Reekie, L.; Jauncey, I.; Payne, D. Low-noise erbium-doped fibre amplifier operating at 1.54 µm. *Electron. Lett.* **1987**, *23*, 1026–1028. [CrossRef]

87. Desurvire, E.; Simpson, J.R.; Becker, P.C. High-gain erbium-doped traveling-wave fiber amplifier. *Opt. Lett.* **1987**, *12*, 888–890. [CrossRef]

88. Miniscalco, W.J. Erbium-doped glasses for fiber amplifiers at 1500 nm. *J. Light. Technol.* **1991**, *9*, 234–250. [CrossRef]

89. Feuchter, T.; Mwarania, E.K.; Wang, J.; Reekie, L.; Wilkinson, J.S. Erbium-doped ion-exchanged waveguide lasers in BK-7 glass. *IEEE Photonics Technol. Lett.* **1992**, *4*, 542–544. [CrossRef]

90. Kitagawa, T.; Hattori, K.; Shuto, K.; Yasu, M.; Kobayashi, M.; Horiguchi, M. Amplification in Erbium-doped silica-based planar lightwave circuits. Optical Amplifiers and Their Applications. Optical Society of America. *Electron. Lett.* **1992**. [CrossRef]

91. Yan, Y.C.; Faber, A.J.; de Waal, H.; Kik, P.G.; Polman, A. Erbium-doped phosphate glass waveguide on silicon with 4.1 dB/cm gain at 1.535 µm. *Appl. Phys. Lett.* **1997**, *71*, 2922–2924. [CrossRef]

92. Polman, A. Erbium implanted thin film photonic materials. *J. Appl. Phys.* **1997**, *82*, 1–39. [CrossRef]

93. Jansen, R.; LaBorde, P.; Lerminiaux, C.; Benveniste, C.; Hall, D. Integrated Optical Signal Amplifier. U.S. Patent N. 5,128,801, 7 July 1992.

94. LaBorde, P. Optical Signal Amplifier Glasses. U.S. Patent N. 5,475,528, 6 March 1995.

95. Ohtsuki, T.; Honkanen, S.; Ingenhoff, J.; Heyden, J.; Fabricius, N.; Najafi, S.I.; Peyghambarian, N. Polarization-insensitive planar glass waveguide amplifiers by silver ion exchange. In Proceedings of the Conference on Lasers and Electro-Optics, Optical Society of America, Anaheim, CA, USA, 2–7 June 1996; p. CThQ6.

96. Delavaux, J.P.; Granlund, S.; Mizuhara, O.; Tzeng, L.D.; Barbier, D.; Rattay, M.; Andre, F.S.; Kevorkian, A. Integrated optics erbium ytterbium amplifier system in 10 Gb/s fiber transmission experiment. In Proceedings of the European Conference on Optical Communication, Oslo, Norway, 19 September 1996; Volume 1, pp. 123–126.

97. Honkanen, S.; Laine, J.P.; Ohtsuki, T.; Tervonen, A.; Peyghambarian, N. Modeling of Er-doped ion-exchanged glass waveguide amplifiers. In *Rare-Earth-Doped Devices*; Honkanen, S., Ed.; SPIE: Bellingham, WA, USA, 1997; Volume 2996, pp. 103–108. [CrossRef]

98. Barbier, D.; Rattay, M.; Saint Andre, F.; Clauss, G.; Trouillon, M.; Kevorkian, A.; Delavaux, J.P.; Murphy, E. Amplifying four-wavelength combiner, based on erbium/ytterbium-doped waveguide amplifiers and integrated splitters. *IEEE Photonics Technol. Lett.* **1997**, *9*, 315–317. [CrossRef]

99. Hempstead, M. Ion-exchanged glass waveguide lasers and amplifiers. In *Rare-Earth-Doped Devices*; Honkanen, S., Ed.; SPIE: Bellingham, WA, USA, 1997; Volume 2996, pp. 94–102. [CrossRef]

100. Shooshtari, A.; Meshkinfam, P.; Touam, T.; Andrews, M.P.; Najafi, S.I. Ion-exchanged Er/Yb phosphate glass waveguide amplifiers and lasers. *Opt. Eng.* **1998**, *37*, 1188–1192. [CrossRef]

101. Man, S.Q.; Wong, R.S.F.; Pun, E.Y.B.; Chung, P.S. Frequency upconversion in Er^{3+}-doped alkali bismuth gallate glasses. In Proceedings of the 1999 IEEE LEOS Annual Meeting Conference, (Cat. No.99CH37009), San Francisco, CA, USA, 8–11 November 1999; Volume 2, pp. 812–813. [CrossRef]

102. Peters, P.M.; Funk, D.S.; Peskin, A.P.; Veasey, D.L.; Sanford, N.A.; Houde-Walter, S.N.; Hayden, J.S. Ion-exchanged waveguide lasers in Er^{3+}-Yb^{3+} codoped silicate glass. *Appl. Opt.* **1999**, *38*, 6879–6886. [CrossRef]

103. Iiyama, K.; Hongo, K.; Demura, F.; Takamiya, S. Erbium/ytterbium co-doped optical waveguide amplifier in soda-lime glass by silver ion exchange. In Proceedings of the Technical Digest, CLEO/Pacific Rim '99, (Cat. No.99TH8464), Seoul, Korea, 30 August–3 September 1999; Volume 4, pp. 1087–1088. [CrossRef]

104. Conti, G.N.; Ayras, P.; Cavaliere, C.; Hwang, B.C.; Luo, T.; Rantala, J.T.; Jiang, S.; Honkanen, S.; Peyghambarian, N. Strip-loaded structure for ion-exchanged Er^{3+}-doped glass waveguide amplifiers. In *Rare-Earth-Doped Materials and Devices III*; Jiang, S., Honkanen, S., Eds.; SPIE: Bellingham, WA, USA, 1999; Volume 3622, pp. 122–128. [CrossRef]

105. Araci, I.E.; Mendes, S.B.; Yurt, N.; Honkanen, S.; Peyghambarian, N. Highly sensitive spectroscopic detection of heme-protein submonolayer films by channel integrated optical waveguide. *Opt. Express* **2007**, *15*, 5595–5603. [CrossRef]

106. Veasey, D.L.; Funk, D.S.; Sanford, N.A.; Hayden, J.S. Arrays of distributed-Bragg-reflector waveguide lasers at 1536 nm in Yb/Er codoped phosphate glass. *Appl. Phys. Lett.* **1999**, *74*, 789–791. [CrossRef]

107. Shmulovich, J.; Muehlner, D.; Bruce, A.; Delavaux, J.M.; McIntosh, C.; Lenz, G.; Gomez, L.; Laskowski, E.; Paunescu, A.; Pafchek, R.; et al. Erbium-doped planar waveguide amplifiers integrated with silica waveguide technology. In Proceedings of the Optical Fiber Communication Conference, Baltimore, MD, USA, 7 March 2000; p. WA1.

108. Hehlen, M.P.; Cockroft, N.J.; Gosnell, T.R.; Bruce, A.J. Spectroscopic properties of Er^{3+}- and Yb^{3+}-doped soda-lime silicate and aluminosilicate glasses. *Phys. Rev. B* **1997**, *56*, 9302–9318. [CrossRef]

109. Vossler, G.L.; Brooks, C.J.; Winik, K.A. Planar Er:Yb glass ion exchanged waveguide laser. *Electron. Lett.* **1995**, *31*, 1162–1163. [CrossRef]

110. Fournier, P.; Meshkinfam, P.; Fardad, M.A.; Andrews, M.P.; Najafi, S.I. Potassium ion-exchanged Er-Yb doped phosphate glass amplifier. *Electron. Lett.* **1997**, *33*, 293–295. [CrossRef]

111. Camy, P.; Roman, J.E.; Willems, F.W.; Hempstead, M.; Van Der Plaats, J.C.; Prel, C.; Beguin, A.; Koonen, A.M.J.; Wilkinson, J.S.; Lerminiaux, C. Ion-exchanged planar lossless splitter at 1.5 μm. *Electron. Lett.* **1996**, *32*, 321–323. [CrossRef]

112. Barbier, D.; Delavaux, J.M.; Kevorkian, A.; Gastaldo, P.; Jouanno, J.M. Yb/Er Integrated optics amplifiers on phosphate glass in single and double pass configurations. In *Optical Fiber Communications Conference*; Optical Society of America: Eindhoven, The Netherlands, 1995; p. PD3. [CrossRef]

113. Barbier, D.; Bruno, P.; Cassagnettes, C.; Trouillon, M.; Hyde, R.L.; Kevorkian, A.; Delavaux, J.M.P. Net gain of 27 dB with a 8.6-cm-long Er/Yb-doped glass-planar-amplifier. In Proceedings of the OFC '98. Optical Fiber Communication Conference and Exhibit, OSA Technical Digest Series Volume2 (IEEE Cat. No.98CH36177), San Jose, CA, USA, 22–27 Febuary 1998; pp. 45–46. [CrossRef]

114. Philipsen, J.L.; Barbier, D.; Kévorkian, A.; Cassagnettes, C.; Krebs, N.; Bruno, P. Compact gain-block consisting of an Er^{3+}-doped waveguide amplifier (EDWA) and a pump/signal multiplexer, realized by ion exchange. In *Optical Amplifiers and Their Applications*; OSA: Québec, QC, Canada, 2000; p. OTuD2. [CrossRef]

115. Martucci, A.; Guglielmi, M.; Fick, J.; Pelli, S.; Forastiere, M.A.; Righini, G.C.; Battaglin, C. Germania sol-gel waveguides for optical amplifiers. In *Sol-Gel Optics V*; Dunn, B.S., Pope, E.J.A., Schmidt, H.K., Yamane, M., Eds.; SPIE: Bellingham, WA, USA, 2000; Volume 3943, pp. 2–9. [CrossRef]

116. Pelli, S.; Brenci, M.; Fossi, M.; Righini, G.C.; Duverger, C.; Montagna, M.; Rolli, R.; Ferrari, M. Optical and spectroscopic characterization of Er/Yb-activated planar waveguides. In *Rare-Earth-Doped Materials and Devices IV*; Jiang, S., Ed.; SPIE: Bellingham, WA, USA, 2000; Volume 3942, pp. 139–145. [CrossRef]

117. Righini, G.C.; Pelli, S.; Fossi, M.; Brenci, M.; Lipovskii, A.A.; Kolobkova, E.V.; Speghini, A.; Bettinelli, M. Characterization of Er-doped sodium-niobium phosphate glasses. In *Rare-Earth-Doped Materials and Devices V*; Jiang, S., Ed.; SPIE: Bellingham, WA, USA, 2001; Volume 4282, pp. 210–215. [CrossRef]

118. Sorbello, G.; Taccheo, S.; Marano, M.; Marangoni, M.; Osellame, R.; Ramponi, R.; Laporta, P. Comparative study of Ag–Na thermal and field-assisted ion exchange on Er-doped phosphate glass. *Opt. Mater.* **2001**, *17*, 425–435. [CrossRef]

119. Kolobkobal, E.V.; Lipovskii, A.A.; Montero, C.; Liñares, J. Formation and modelling of optically waveguiding structures in a high-concentration Er-doped phosphate glass *J. Phys. D Appl. Phys.* **1999**, *32*, L9. [CrossRef]

120. Lin, H.; Jiang, S.; Wu, J.; Song, F.; Peyghambarian, N.; Pun, E.Y.B. Er^{3+}-doped Na_2O–Nb_2O_5–TeO_2 glasses for optical waveguide laser and amplifier. *J. Phys. Appl. Phys.* **2003**, *36*, 812–817. [CrossRef]

121. Conti, G.N.; Tikhomirov, V.K.; Bettinelli, M.; Berneschi, S.; Brenci, M.; Chen, B.; Pelli, S.; Speghini, A.; Seddon, A.B.; Righini, G.C. Characterization of ion-exchanged waveguides in tungsten tellurite and zinc tellurite Er3+-doped glasses. *Opt. Eng.* **2003**, *42*, 2805–2811. [CrossRef]

122. Pelli, S.; Bettinelli, M.; Brenci, M.; Calzolai, R.; Chiasera, A.; Ferrari, M.; Nunzi Conti, G.; Speghini, A.; Zampedri, L.; Zheng, J.; et al. Erbium-doped silicate glasses for integrated optical amplifiers and lasers. *J. Non-Cryst. Solids* **2004**, *345–346*, 372–376. [CrossRef]

123. Liu, K.; Pun, E.Y.B. K^+-Na^+ ion-exchanged waveguides in Er^{3+}-Yb^{3+} codoped phosphate glasses using field-assisted annealing. *Appl. Opt.* **2004**, *43*, 3179–3184. [CrossRef] [PubMed]

124. Pissadakis, S.; Ikiades, A.; Hua, P.; Sheridan, A.K.; Wilkinson, J.S. Photosensitivity of ion-exchanged Er-doped phosphate glass using 248 nm excimer laser radiation. *Opt. Express* **2004**, *12*, 3131–3136. [CrossRef] [PubMed]

125. Capek, P.; Mika, M.; Oswald, J.; Tresnakova, P.; Salavcova, L.; Kolek, O.; Schrofel, J.; Spirkova, J. Effect of divalent cations on properties of Er^{3+}-doped silicate glasses. *Opt. Mater.* **2004**, *27*, 331–336. [CrossRef]

126. Mika, M.; Kolek, O.; Spirkova, J.; Capek, P.; Berneschi, S.; Brenci, M.; Conti, G.N.; Pelli, S.; Sebastiani, S.; Righini, G.C. The effect of Ca^{2+}, Mg^{2+}, and Zn^{2+} on optical properties of Er^{3+}-doped silicate glass. In *Optical Components and Materials II*; Jiang, S., Digonnet, M.J., Eds.; SPIE: Bellingham, WA, USA, 2005; Volume 5723, pp. 63–70. [CrossRef]

127. Ondrac, F.; Ek; Salavcova, L.; Míka, M.; Špirková, J. Characterization of Erbium Doped Glass Optical Waveguides by a Fine Tunable Semiconductor Laser. *Meas. Sci. Rev.* **2005**, *5*, 6–9.

128. Bucci, D.; Grelin, J.; Ghibaudo, E.; Broquin, J.E. Study of a pump/signal multiplexer based on a segmented asymmetric Y junction by silver/sodium ion exchange on glass. *Proc. SPIE* **2006**, *6123*, 246–254. [CrossRef]

129. Berneschi, S.; Bettinelli, M.; Brenci, M.; Dall'Igna, R.; Nunzi Conti, G.; Pelli, S.; Profilo, B.; Sebastiani, S.; Speghini, A.; Righini, G.C. Optical and spectroscopic properties of soda-lime alumino silicate glasses doped with Er^{3+} +and/or Yb^{3+}. *Opt. Mater.* **2006**, *28*, 1271–1275. [CrossRef]

130. Yliniemi, S.; Albert, J.; Laronche, A.; Castro, J.M.; Geraghty, D.; Honkanen, S. Negligible birefringence in dual-mode ion-exchanged glass waveguide gratings. *Appl. Opt.* **2006**, *45*, 6602–6606. [CrossRef]

131. Nandi, P.; Jose, G. Erbium doped phospho-tellurite glasses for 1.5 µm optical amplifiers. *Opt. Commun.* **2006**, *265*, 588–593. [CrossRef]

132. Rivera, V.; Chillcce, E.; Rodriguez, E.; Cesar, C.; Barbosa, L. Planar waveguides by ion exchange in Er^{3+}-doped tellurite glass. *J. Non-Cryst. Solids* **2006**, *352*, 363–367. [CrossRef]

133. Sakida, S.; Nanba, T.; Miura, Y. Refractive-index profiles and propagation losses of Tb^{3+}-doped tungsten tellurite glass waveguide by Ag+–Na+ ion-exchange. *Mater. Lett.* **2006**, *60*, 3413–3415. [CrossRef]

134. Bucci, D.; Grelin, J.; Ghibaudo, E.; Broquin, J. Realization of a 980-nm/1550-nm Pump-Signal (De)multiplexer Made by Ion-Exchange on Glass Using a Segmented Asymmetric Y-Junction. *IEEE Photonics Technol. Lett.* **2007**, *19*, 698–700. [CrossRef]

135. Ondráček, F.; Salavcová, L.; Míka, M.; Lahodný, F.; Slavík, R.; Špirková, J.; Čtyroký, J. Fabrication and characterization of channel optical waveguides in Er-Yb-doped silicate glasses. *Opt. Mater.* **2007**, *30*, 457–461. [CrossRef]

136. Sakida, S.; Nanba, T.; Miura, Y. Optical properties of Er^{3+}-doped tungsten tellurite glass waveguides by Ag-Na ion-exchange. *Opt. Mater.* **2007**, *30*, 586–593. [CrossRef]

137. Svecova, B.; Spirkova, J.; Janakova, S.; Mika, M. Ion-exchanged optical waveguides fabricated in novel Er^{3+} and Er^{3+}. Yb^{3+}-doped silicate glasses: Relations between glass composition, basicity and waveguide properties. *Mater. Sci. Eng. B* **2008**, *149*, 177–180. [CrossRef]

138. Barbosa, A.; Maia, L.; Nascimento, A.; Gonçalves, R.; Poirier, G.; Messaddeq, Y.; Ribeiro, S. Er^{3+}-doped germanate glasses for active waveguides prepared by Ag or K - Na ion-exchange. *J. Non-Cryst. Solids* **2008**, *354*, 4743–4748. [CrossRef]

139. Svecova, B.; Spirkova, J.; Janakova, S.; Mika, M.; Oswald, J.; Mackova, A. Diffusion process applied in fabrication of ion-exchanged optical waveguides in novel Er^{3+} and Er^{3+} - Yb^{3+}-doped silicate glasses. *J. Mater. Sci. Mater. Electron.* **2009**, *20*, 510–513. [CrossRef]

140. Yang, D.L.; Pun, E.Y.; Lin, H. Tm^{3+}-doped ion-exchanged aluminum germanate glass waveguide for S-band amplification. *Appl. Phys. Lett.* **2009**, *95*, 151106. [CrossRef]

141. Bozelli, J.C.; Nunes, L.A.d.O.; Sigoli, F.A.; Mazali, I.O. Erbium and Ytterbium Codoped Titanoniobophosphate Glasses for Ion-Exchange-Based Planar Waveguides. *J. Am. Ceram. Soc.* **2010**, *93*, 2689–2692. [CrossRef]

142. Kimura, K.; Sakida, S.; Benino, Y.; Nanba, T. Fabrication and characterization of Er^{3+}-doped tellurite glass waveguide by Ag-Na ion-exchange method. *IOP Conf. Ser. Mater. Sci. Eng.* **2011**, *18*, 112018. [CrossRef]

143. Olivier, M.; Doualan, J.L.; Camy, P.; Lhermite, H.; Adam, J.L.; Nazabal, V. Development of Praseodymium doped fluoride waveguide. In *Optical Components and Materials IX*; Jiang, S., Digonnet, M.J.F., Dries, J.C., Eds.; SPIE: Bellingham, WA, USA, 2012; Volume 8257, pp. 26–32. [CrossRef]

144. Shen, L.; Chen, B.; Lin, H.; Pun, E. Praseodymium ion doped phosphate glasses for integrated broadband ion-exchanged waveguide amplifier. *J. Alloys Compd.* **2015**, *622*, 1093–1097. [CrossRef]

145. Wong, S.F.; Pun, E.Y.B.; Chung, P.S. Er^{3+}-Yb^{3+} codoped phosphate glass waveguide amplifier using Ag-Li ion exchange. *IEEE Photonics Technol. Lett.* **2002**, *14*, 80–82. [CrossRef]

146. Chen, H.Y.; Liu, Y.Z.; Dai, J.Z.; Yang, Y.P.; Guan, Z.G.; Huang, X.L. Er^{3+}/Yb^{3+} codoped phosphate glass waveguide amplifier. In Proceedings of the IEEE 2002 International Conference on Communications, Circuits and Systems, Chengdu, China, 29 June–1 July 2002; Volume 1, pp. 827–829. [CrossRef]

147. Jose, G.; Sorbello, G.; Taccheo, S.; Cianci, E.; Foglietti, V.; Laporta, P. Active waveguide devices by Ag–Na ion exchange on erbium–ytterbium doped phosphate glasses. *J. Non-Cryst. Solids* **2003**, *322*, 256–261. [CrossRef]

148. Gardillou, F.; Bastard, L.; Broquin, J.E. 4.25 dB gain in a hybrid silicate/phosphate glasses optical amplifier made by wafer bonding and ion-exchange techniques. *Appl. Phys. Lett.* **2004**, *85*, 5176–5178. [CrossRef]

149. Gardillou, F.; Broquin, J.E. Optical amplifier made by reporting an Er^{3+}-Yb^{3+}-codoped glass layer on an ion-exchanged passive glass substrate by wafer bonding. In *Integrated Optics: Devices, Materials, and Technologies IX*; Sidorin, Y., Waechter, C.A., Eds.; SPIE: Bellingham, WA, USA, 2005; Volume 5728, pp. 120–128. [CrossRef]

150. Guo-Liang, J.; Gong-Wang, S.; Huan, M.; Li-Li, H.; Qu, L. Gain and Noise Figure of a Double-Pass Waveguide Amplifier Based on Er/Yb-Doped Phosphate Glass. *Chin. Phys. Lett.* **2005**, *22*, 2862–2864. [CrossRef]

151. Gardillou, F.; Broquin, J.E. Net gain demonstration with glass hybrid optical amplifiers made by ion-exchange and wafer bonding. *Integr. Opt. Devices Mater. Technol. X* **2006**, *2163*. [CrossRef]

152. Della Valle, G.; Taccheo, S.; Laporta, P.; Sorbello, G.; Cianci, E.; Foglietti, V. Compact high gain erbium-ytterbium doped waveguide amplifier fabricated by Ag-Na ion exchange. *Electron. Lett.* **2006**, *42*, 632–633. [CrossRef]

153. Zian, H.; Yigang, L.; Yanwu, Z.; Dongxiao, L.; Liying, L.; Lei, X. Er^{3+}-Yb^{3+} co-doped waveguide amplifier and lossless power splitter fabricated by a two-step ion exchange on a commercial phosphate glass. *J. Korean Phys. Soc.* **2006**, *49*, 2159–2163.

154. Zhang, X.; Liu, K.; Mu, S.; Tan, C.; Zhang, D.; Pun, E.; Zhang, D. Er^{3+}–Yb^{3+} co-doped glass waveguide amplifiers using ion exchange and field-assisted annealing. *Opt. Commun.* **2006**, *268*, 300–304. [CrossRef]

155. Liu, K.; Pun, E.Y. Modeling and experiments of packaged Er^{3+}–Yb^{3+} co-doped glass waveguide amplifiers. *Opt. Commun.* **2007**, *273*, 413–420. [CrossRef]

156. Ondracek, F.; Jagerska, J.; Salavcova, L.; Mika, M.; Spirkova, J.; Ctyroky, J. Er–Yb Waveguide Amplifiers in Novel Silicate Glasses. *IEEE J. Quantum Electron.* **2008**, *44*, 536–541. [CrossRef]

157. Shao, G.; Jin, G.; Li, Q. Gain and noise figure characteristics of an Er^{3+}–Yb^{3+} doped phosphate glass waveguide amplifier with a bidirectional pump scheme and double-pass configuration. *Opt. Eng.* **2008**, *47*, 1–7. [CrossRef]

158. Onestas, L.; Nappez, T.; Ghibaudo, E.; Vitrant, G.; Broquin, J.E. 980 nm–1550 nm vertically integrated duplexer for hybrid erbium-doped waveguide amplifiers on glass. In *Integrated Optics: Devices, Materials, and Technologies XIII*; Broquin, J.E., Greiner, C.M., Eds.; SPIE: Bellingham, WA, USA, 2009; Volume 7218, pp. 40–52. [CrossRef]

159. Shao, G.; Jin, G.; Zhan, L.; Li, Q. Influence of multimode-pump propagation on the gain characteristics of erbium-ytterbium doped waveguide amplifier. *Appl. Phys. B* **2009**, *97*, 67–71. [CrossRef]

160. Donzella, V.; Toccafondo, V.; Faralli, S.; Pasquale, F.D.; Cassagnettes, C.; Barbier, D.; Figueroa, H.H. Ion-exchanged Er^{3+}/Yb^{3+} co-doped waveguide amplifiers longitudinally pumped by broad area lasers. *Opt. Express* **2010**, *18*, 12690–12701. [CrossRef]

161. Gong, H.; Lin, L.; Zhao, X.; Pun, E.; Yang, D.; Lin, H. Mixing up-conversion excitation behaviors in Tb^{3+}–Tb^{3+} codoped aluminum germanate glasses for visible waveguide devices. *J. Alloys Compd.* **2010**, *503*, 133–137. [CrossRef]

162. Onestas, L.; Bucci, D.; Ghibaudo, E.; Broquin, J. Vertically Integrated Broadband Duplexer for Erbium-Doped Waveguide Amplifiers Made by Ion Exchange on Glass. *IEEE Photonics Technol. Lett.* **2011**, *23*, 648–650. [CrossRef]

163. He, Z.; Li, Y.; Li, Y.; Zhang, Y.; Liu, L.; Xu, L. Ion-exchanged silica-on-silicon structured channel erbium-doped waveguide amplifiers. *Appl. Opt.* **2011**, *50*, 2964–2972. [CrossRef]

164. Šmejcký, J.; Jeřábek, V.; Nekvindová, P. Gain determination of optical active doped planar waveguides. In *Photonics, Devices, and Systems VII*; Fliegel, K., Páta, P., Eds.; SPIE: Bellingham, WA, USA, 2017; Volume 10603, pp. 152–162. [CrossRef]

165. Šmejcký, J.; Jeřábek, V. Differential Gain Comparison of Optical Planar Amplifier on Silica Glasses Doped with Bi-Ge and Er, Yb Ions. *J. Mater. Sci. Technol. Res.* **2020**, *7*, 71–79.

166. Yliniemi, S.; Honkanen, S.; Ianoul, A.; Laronche, A.; Albert, J. Photosensitivity and volume gratings in phosphate glasses for rare-earth-doped ion-exchanged optical waveguide lasers. *J. Opt. Soc. Am. B* **2006**, *23*, 2470–2478. [CrossRef]

167. Salas-Montiel, R.; Bastard, L.; Grosa, G.; Broquin, J.E. Hybrid Neodymium-doped passively Q-switched waveguide laser. *Mater. Sci. Eng. B* **2008**, *149*, 181–184. [CrossRef]

168. Casale, M.; Bucci, D.; Bastard, L.; Broquin, J.E. 1.55 μm hybrid waveguide laser made by ion-exchange and wafer bonding. In *Integrated Optics: Devices, Materials, and Technologies XVI*; Broquin, J.E., Conti, G.N., Eds.; SPIE: Bellingham, WA, USA, 2012; Volume 8264, pp. 32–39. [CrossRef]

169. Choudhary, A.; Dhingra, S.; D'Urso, B.; Kannan, P.; Shepherd, D.P. Graphene Q-Switched Mode-Locked and Q-Switched Ion-Exchanged Waveguide Lasers. *IEEE Photonics Technol. Lett.* **2015**, *27*, 646–649. [CrossRef]

170. Ding, Y.; Jiang, S.; Luo, T.; Hu, Y.; Peyghambarian, N. Optical waveguides prepared in Er3+-doped tellurite glass by Ag+-Na+ ion exchange. In *Rare-Earth-Doped Materials and Devices V*; Jiang, S., Ed.; SPIE: Bellingham, WA, USA, 2001; Volume 4282, pp. 23–30. [CrossRef]

171. Conti, G.N.; Berneschi, S.; Bettinelli, M.; Brenci, M.; Chen, B.; Pelli, S.; Speghini, A.; Righini, G. Rare-earth doped tungsten tellurite glasses and waveguides: Fabrication and characterization. *J. Non-Cryst. Solids* **2004**, *345–346*, 343–348. [CrossRef]

172. Zhao, S.; Wang, X.; Fang, D.; Xu, S.; Hu, L. Spectroscopic properties and thermal stability of Er^{3+}-doped tungsten tellurite glass for waveguide amplifier application. *J. Alloys Compd.* **2006**, *424*, 243–246. [CrossRef]

173. Righini, G.C.; Forastiere, M.A.; Guglielmi, M.; Martucci, A. Rare-earth-doped sol-gel waveguides: A review. In *Rare-Earth-Doped Devices II*; Honkanen, S., Jiang, S., Eds.; SPIE: Bellingham, WA, USA, 1998; Volume 3280, pp. 57–66. [CrossRef]

174. Huang, W.; Syms, R.R.A.; Yeatman, E.M.; Ahmad, M.M.; Clapp, T.V.; Ojha, S.M. Fiber-device-fiber gain from a sol-gel erbium-doped waveguide amplifier. *IEEE Photonics Technol. Lett.* **2002**, *14*, 959–961. [CrossRef]

175. Peled, A.; Chiasera, A.; Nathan, M.; Ferrari, M.; Ruschin, S. Monolithic rare-earth doped sol-gel tapered rib waveguide laser. *Appl. Phys. Lett.* **2008**, *92*, 221104. [CrossRef]

176. Aquino, F.T.; Ferrari, J.L.; Ribeiro, S.J.L.; Ferrier, A.; Goldner, P.; Gonçalves, R.R. Broadband NIR emission in novel sol–gel Er^{3+}-doped SiO_2–Nb_2O_5 glass ceramic planar waveguides for photonic applications. *Opt. Mater.* **2013**, *35*, 387–396. [CrossRef]

177. Almeida, R.M.; Marques, A.C. The potential of ion exchange in sol–gel derived photonic materials and structures. *Mater. Sci. Eng. B* **2008**, *149*, 118–122. [CrossRef]

178. Jaouen, Y.; du Mouza, L.; Barbier, D.; Delavaux, J..; Bruno, P. Eight-wavelength Er-Yb doped amplifier: Combiner/splitter planar integrated module. *IEEE Photonics Technol. Lett.* **1999**, *11*, 1105–1107. [CrossRef]

179. Chen, Q.; Milanese, D.; Chen, Q.; Ferraris, M.; Righini, G.C. Fabrication and direct bonding of photosensitive multicomponent silicate glasses for lossless planar waveguide splitters. *J. Non-Cryst. Solids* **2008**, *354*, 1230–1234. [CrossRef]

180. Bradley, J.; Pollnau, M. Erbium-doped integrated waveguide amplifiers and lasers. *Laser Photonics Rev.* **2011**, *5*, 368–403. [CrossRef]

181. Enrichi, F.; Cattaruzza, E.; Ferrari, M.; Gonella, F.; Martucci, A.; Ottini, R.; Riello, P.; Righini, G.C.; Trave, E.; Vomiero, A.; et al. Role of Ag multimers as broadband sensitizers in Tb3+/Yb3+ co-doped glass-ceramics. In *Fiber Lasers and Glass Photonics*; Taccheo, S., Mackenzie, J.I., Ferrari, M., Eds.; SPIE: Bellingham, WA, USA, 2018; Volume 10683, pp. 139–147. [CrossRef]

182. Vařák, P.; Vytykáčová, S.; Nekvindová, P.; Michalcová, A.; Malinský, P. The influence of copper and silver in various oxidation states on the photoluminescence of Ho3+/Yb3+ doped zinc-silicate glasses. *Opt. Mater.* **2019**, *91*, 253–260. [CrossRef]

183. Enrichi, F.; Belmokhtar, S.; Benedetti, A.; Bouajaj, A.; Cattaruzza, E.; Coccetti, F.; Colusso, E.; Ferrari, M.; Ghamgosar, P.; Gonella, F.; et al. Ag nanoaggregates as efficient broadband sensitizers for Tb^{3+} ions in silica-zirconia ion-exchanged sol-gel glasses and glass-ceramics. *Opt. Mater.* **2018**, *84*, 668–674. [CrossRef]

184. Vařák, P.; Nekvindová, P.; Vytykáčová, S.; Michalcová, A.; Malinský, P.; Oswald, J. Near-infrared photoluminescence enhancement and radiative energy transfer in RE-doped zinc-silicate glass (RE = Ho, Er, Tm) after silver ion exchange. *J. Non-Cryst. Solids* **2021**, *557*, 120580. [CrossRef]

185. Li, N. Rare-Earth-Doped Lasers on Silicon Photonics Platforms. Ph.D. Thesis, Harvard University, Cambridge, MA, USA, 2018.

186. Demirtaş, M.; Ay, F. High-Gain Er^{3+}:Al_2O_3 On-Chip Waveguide Amplifiers. *IEEE J. Sel. Top. Quantum Electron.* **2020**, *26*, 1–8. [CrossRef]

187. Hendriks, W.A.P.M.; Chang, L.; van Emmerik, C.I.; Mu, J.; de Goede, M.; Dijkstra, M.; Garcia-Blanco, S.M. Rare-earth ion doped Al_2O_3 for active integrated photonics. *Adv. Phys. X* **2021**, *6*, 1833753. [CrossRef]

188. Lipovskii, A.; Zhurikhina, V.; Tagantsev, D. 2D-structuring of glasses via thermal poling: A short review. *Int. J. Appl. Glass Sci.* **2018**, *9*, 24–28. [CrossRef]

189. Mateo, E.F.; Linares, J. A phase insensitive all-optical router based on nonlinear lenslike planar waveguides. *Opt. Express* **2005**, *13*, 3355–3370. [CrossRef]

190. Gerasimenko, V.S.; Gerasimenko, N.D.; Kiselev, F.D. Numerical modelling of an error of manufacturing of ion- exchange waveguide for the tasks of quantum computations. *J. Phys. Conf. Ser.* **2019**, *1410*, 012136-1-5. [CrossRef]

191. Gerasimenko, V.S.; Gerasimenko, N.D.; Kiselev, F.D.; Samsonov, E.; Kozlov, S. Numerical modeling of ion exchange waveguide for the tasks of quantum computations. *Nanosyst. Phys. Chem. Math* **2019**, *10*, 147–153. [CrossRef]

192. Nahra, S.P.M.; Joos, M.; Muhammad, M.H.; Agafonov, V.; Lhuillier, E.; Geoffra, F.; Davydov, V.; Glorieu, Q.; Giacobin, E.; Blaiz, S.; et al. Nanophotonic approaches for integrated quantum photonics. *arXiv* **2019**, arXiv:1909.10343.

193. Couteau, C.; Nahra, M.; Muhammad, M.H.; Pierini, S.; Xu, X.; Broussier, A.; Bachelot, R.; Blaize, S. Towards a new platform for quantum photonics applications. In *Advances in Photonics of Quantum Computing, Memory, and Communication XII*; Hemmer, P.R., Migdall, A.L., Hasan, Z.U., Eds.; SPIE: Bellingham, WA, USA, 2019; Volume 10933, pp. 88–90. [CrossRef]

194. Prieto-Blanco, X.; Montero-Orille, C.; Liñares, J.; González-Núñez, H.; Balado, D. Quantum projectors implemented with optical directional couplers fabricated by Na/K ion-exchange in soda-lime glass. *arXiv* **2021**, arXiv:2102.01169.

195. Schröder, H.; Neitz, M.; Schneider-Ramelow, M. Demonstration of glass-based photonic interposer for mid-board-optical engines and electrical-optical circuit board (EOCB) integration strategy. In *Optical Interconnects XVIII*; Schröder, H., Chen, R.T., Eds.; SPIE: Bellingham, WA, USA, 2018; Volume 10538, pp. 57–69. [CrossRef]

196. Spring, J.; Metcalf, B.; Humphreys, P.; Kolthammer, W.; Jin, X.; Barbieri, M.; Datta, A.; Thomas-Peter, N.; Langford, N.K.; Kundys, D.; et al. Boson sampling on photonic chip. *Science* **2013**, *239*, 798–801. [CrossRef]

197. Perets, H.B.; Lahini, Y.; Pozzi, F.; Sorel, M.; Morandotti, R.; Silberberg, Y. Realization of Quantum Walks with Negligible Decoherence in Waveguide Lattices. *Phys. Rev. Lett* **2008**, *100*, 170506. [CrossRef]

198. Balado, D.; Prieto-Blanco, X.; Barral, D.; Liñares, J. Autocompensating high-dimensional quantum cryptography by using integrated photonic devices in multicore optical fiber spatial multiplexing systems. In Proceedings of the International Conference on Integrated Quantum Photonics (ICIQP), Paris, France, 15–17 October 2018; p. 75.

199. Yip, G.L.; Finak, J. Directional-coupler power divider by two-step K^+-ion exchange. *Opt. Lett.* **1984**, *9*, 423–425. [CrossRef]

200. Honkanen, S.; Pöyhönen, P.; Tervonen, A.; Najafi, S.I. Waveguide coupler for potassium- and silver-ion-exchanged waveguides in glass. *Appl. Opt.* **1993**, *32*, 2109–2111. [CrossRef]

201. Righini, G.C.; Conti, G.N.; Forastiere, M.A. Integrated optical directional couplers: How effective are design and modeling for device production? In *Integrated Optics Devices: Potential for Commercialization*; Najafi, S.I., Armenise, M.N., Eds.; SPIE: Bellingham, WA, USA, 1997; Volume 2997, pp. 212–219. [CrossRef]

202. Li, G.; Bai, N.; Zhao, N.; Xia, C. Space-division multiplexing: The next frontier in optical communication. *Adv. Opt. Photon.* **2014**, *6*, 413–487. [CrossRef]

203. Pauwels, J.; Van der Sande, G.; Verschaffelt, G. Space division multiplexing in standard multi-mode optical fibers based on speckle pattern classification. *Sci. Rep.* **2019**, *9*, 17597. [CrossRef]

204. Montero-Orille, C.; Moreno, V.; Prieto-Blanco, X.; Mateo, E.F.; Ip, E.; Crespo, J.; Liñares, J. Ion-exchanged glass binary phase plates for mode-division multiplexing. *Appl. Opt.* **2013**, *52*, 2332–2339. [CrossRef]

205. Prieto-Blanco, X.; Montero-Orille, C.; Moreno, V.; Mateo, E.F.; Liñares, J. Chromatic characterization of ion-exchanged glass binary phase plates for mode-division multiplexing. *Appl. Opt.* **2015**, *54*, 3308–3314. [CrossRef]

206. Ip, E.; Milione, G.; Li, M.J.; Cvijetic, N.; Kanonakis, K.; Stone, J.; Peng, G.; Prieto, X.; Montero, C.; Moreno, V.; et al. SDM transmission of real-time 10GbE traffic using commercial SFP+ transceivers over 0.5km elliptical-core few-mode fiber. *Opt. Express* **2015**, *23*, 17120–17126. [CrossRef]

207. Liñares, J.; Montero-Orille, C.; Moreno, V.; Mouriz, D.; Nistal, M.C.; Prieto-Blanco, X. Ion-exchanged glass binary phase plates for mode multiplexing in graded-index optical fibers. *Appl. Opt.* **2017**, *56*, 7099–7106. [CrossRef]

208. Liñares, J.; Prieto-Blanco, X.; Moreno, V.; Montero-Orille, C.; Mouriz, D.; Nistal, M.C.; Barral, D. Interferometric space-mode multiplexing based on binary phase plates and refractive phase shifters. *Opt. Express* **2017**, *25*, 10925–10938. [CrossRef]

209. Liñares, J.; Prieto-Blanco, X.; Montero-Orille, C.; Moreno, V. Spatial mode multiplexing/demultiplexing by Gouy phase interferometry. *Opt. Lett.* **2017**, *42*, 93–96. [CrossRef]

210. Riedel1, M.F.; Binosi, D.; Thew, R.; Calarco1, T. The European quantum technologies flagship programme. *Quantum Sci. Technol.* **2017**, *2*, 030501. [CrossRef]

211. Gibney, E. Chinese satellite is one giant step for the quantum internet. *Nature* **2016**, *535*, 478–479. [CrossRef]

212. Cavaliere, F.; Prati, E.; Poti, L.; Muhammad, I.; Catuogno, T. Secure Quantum Communication Technologies and Systems: From Labs to Markets. *Quantum Rep.* **2020**, *2*, 80–106. [CrossRef]

213. Integrated Quantum Optical Circuits Market by Material Type and Application: Global Opportunity Analysis and Industry Forecast, 2018–2025. Available online: https://www.alliedmarketresearch.com/integrated-quantum-optical-circuits-market (accessed on 20 May 2021).

214. Bennett, C.H.; Brassard, G. Quantum cryptography: Public key distribution and coin tossing. In Proceedings of the IEEE International Conference on Computers, Systems and Signal Processing, Bangalore, India, 10–12 December 1984; pp. 175–179.

215. Liñares-Beiras, J.; Prieto-Blanco, X.; Balado, D.; Carral, G.M.. Autocompensating high-dimensional quantum cryptography by phase conjugation in optical fibers. *EPJ Web Conf.* **2020**, *238*, 11004. [CrossRef]

216. Liñares, J.; Prieto-Blanco, X.; Balado, D.; Carral, G.M. Fully autocompensating high-dimensional quantum cryptography by quantum degenerate four-wave mixing. *Phys. Rev. A* **2021**, *103*, 043710. [CrossRef]

217. Lo, H.K.; Curty, M.; Qi, B. Measurement-Device-Independent Quantum Key Distribution. *Phys. Rev. Lett.* **2012**, *108*, 130503. [CrossRef]

218. Xavier, J.; Yu, D.; Jones, C.; Zossimova, E.; Vollmer, F. Quantum nanophotonic and nanoplasmonic sensing: Towards quantum optical bioscience laboratories on chip. *Nanophotonics* **2021**, *10*, 1387–1435. [CrossRef]

219. Janszky, J.; Sibilia, C.; Bertolotti, M.; Yushin, Y. Non-classical Light in a Linear Coupler. *J. Mod. Opt.* **1988**, *35*, 1757–1765. [CrossRef]

220. Perina J., Jr.; Perina, J. Quantum statistics of nonlinear optical couplers (Chapter 5). In *Progress in Optics*; Wolf, E., Ed.; Elsevier: Amsterdam, The Netherlands, 2000; Volume 41, pp. 362–419.

221. Liñares, J.; Nistal, M.C. Quantization of coupled modes propagation in integrated optical waveguides. *J. Mod. Opt.* **2003**, *50*, 781–790. [CrossRef]

222. Liñares, J.; Nistal, M.C.; Barral, D. Quantization of Coupled 1D Vector Modes in Integrated Photonics Waveguides. *New J. Phys* **2008**, *10*, 063023.1–063023.10. [CrossRef]

223. Yuan, J.; Luo, F.; Cao, M.; Chen, W. MMI splitter by ion exchange on K9. In *Passive Components and Fiber-based Devices II*; Sun, Y., Chen, J., Lee, S.B., White, I.H., Eds.; International Society for Optics and Photonics, SPIE: Bellingham, WA, USA, 2005; Volume 6019, pp. 851–857. [CrossRef]

224. Barkman, O.; JeYábek, V.; Prajzler, V. Optical Splitters Based on Self-Imaging Effect in Multi-Mode Waveguide Made by Ion Exchange in Glass. *Radioengineering* **2013**, *22*, 352–356.

225. Gnewuch, H.; Román, J.E.; Hempstead, M.; Wilkinson, J.S.; Ulrich, R. Beat-length measurement in directional couplers by thermo-optic modulation. *Opt. Lett.* **1996**, *21*, 1189–1191. [CrossRef]
226. Stosch, J.H.; Kühler, T.; Griese, E. Optical directional coupler for graded index waveguides in thin glass sheets for PCB integration. In Proceedings of the 2016 IEEE 20th Workshop on Signal and Power Integrity (SPI), Turin, Italy, 8–11 May 2016; pp. 1–4. [CrossRef]
227. Liñares, J.; Montero, C.; Moreno, V.; Nistal, M.C.; Prieto, X.; Salgueiro, J.R.; Sotelo, D. Glass processing by ion exchange to fabricate integrated optical planar components: Applications. *Proc. SPIE* **2000**, *3936*, 227–238. [CrossRef]

Article

A Polygonal Model to Design and Fabricate Ion-Exchanged Diffraction Gratings

Carlos Montero-Orille *, Xesús Prieto-Blanco, Héctor González-Núñez and Jesús Liñares

Quantum Materials and Photonics Research Group, Optics Area, Department of Applied Physics, Faculty of Physics/Faculty of Optics and Optometry, Universidade de Santiago de Compostela, Campus Vida s/n, E-15782 Santiago de Compostela, Galicia, Spain; xesus.prieto.blanco@usc.es (X.P.-B.); hector.gonzalez.nunez@usc.es (H.G.-N.); suso.linares.beiras@usc.es (J.L.)
* Correspondence: carlos.montero@usc.es

Abstract: We propose a simple polygonal model to describe the phase profile of ion-exchanged gratings. This model enables the design of these gratings, as well as the characterization of the ion-exchange process itself. Several ion-exchanged gratings were fabricated to validate the model and to characterize the process involved in their fabrication. From this characterization, we show the practical utility of the model by designing and fabricating both a grating that removes the zero order and a three splitter. The performance of these two elements was good, although the first one stood out especially because only 0.5% of the power remained in the zero order after diffraction. This polygonal model could be useful to design more complex diffractive elements.

Keywords: ion-exchange; phase gratings; diffractive optics

Citation: Montero-Orille, C.; Prieto-Blanco, X.; González-Núñez, H.; Liñares, J. A Polygonal Model to Design and Fabricate Ion-Exchanged Diffraction Gratings. *Appl. Sci.* **2021**, *11*, 1500. https://doi.org/10.3390/app11041500

Academic Editor: Andrey Miroshnichenko

Received: 10 December 2020
Accepted: 1 February 2021
Published: 7 February 2021

Publisher's Note: MDPI stays neutral with regard to jurisdictional claims in published maps and institutional affiliations.

1. Introduction

Diffractive Optical Elements (DOEs) are widely used in different fields such as holography [1], spectroscopy [2], astronomy [3], and so on. Among DOEs, diffraction gratings stand out for their versatility, being a common part of devices for wavelength division multiplexing, optical pulse compression, or laser tuning. Gratings modulate the amplitude or the phase of the light incident on them. As a result, light is diffracted into the so-called diffraction orders. In applications where the diffraction efficiency is important, phase modulation is preferred to amplitude modulation. The most investigated phase gratings are those with sinusoidal profiles. The gratings with this kind of profile have significant advantages. First, they can be mathematically analyzed by closed-form expressions. Secondly, these profiles can give rise to many interesting diffraction patterns including the cancellation of the zero order, splitting in several orders, and so on. Lastly, sinusoidal gratings can be fabricated by interference of two plane waves on photosensitive substrates. However, gratings fabricated in this way are delicate, and specific phase profiles are difficult to obtain. Here, we use the ion-exchange technique to fabricate robust and precise sinusoidal-like phase gratings by taking advantage of its simplicity, low cost, and easy control of the fabrication parameters [4]. Moreover, the flat surface of the elements fabricated by this technique permits straightforward cleaning and antireflection coating, as well as integration in stacked optical systems [5]. As is known, ion-exchange changes the refractive index of suitable substrates, typically in a region of some micrometers under its surface. This change can be made selectively in only some regions of the substrate, provided that a mask for the ions is used. In particular, by designing appropriate masks, as well as the ion-exchange process itself, different kinds of phase gratings have been fabricated [6,7]. In these works, the authors successfully designed and fabricated grating beam splitters by using a numerical approach for both the ion-exchange process and the diffraction calculations. However, the design of such elements using numerical methods is not easy when trying to make complex elements. In the present work, we propose a simple polygonal model for the phase profile achieved following thermal ion-exchange processes. This model enables the design and fabrication of grating elements in a simple way. After characterization of the ion-exchange process by using such a model, different grating elements can be designed and fabricated easily. As examples, we design and fabricate a diffraction grating with suppressed zero order and a three splitter. Cancellation of the zero order of diffraction is desirable in many applications of diffractive elements, such as digital holography [8], spatial light modulation [9], and grating fabrication [10], among others. On the other hand, light splitting is useful, in general, for power division tasks [11], and in particular, it has been used, combined with other phase functions, in the fabrication of vortex gratings [12].

In Section 2, we explain the characteristics of the model and obtain expressions for the transmission function of gratings with such a phase profile. Next, we describe the characterization procedure of the two-dimensional ion-exchange process by using this model. This characterization provides the characteristic parameters of the model as a function of the ion-exchange time. Finally, in Section 4, we check the potential of the model by designing and fabricating a grating with suppressed zero order and a three splitter.

2. Polygonal Model for the Phase Profile of Ion-Exchanged Gratings

The phase profile of a grating is conditioned by its fabrication process. In particular, in the fabrication of gratings by thermal ion-exchange through a mask, the phase profile generated is determined not only by the shape of the mask, but also by the ion-exchange process itself. As a first approach, the phase profile of a grating fabricated by this technique would be a square profile. However, when a thermal ion-exchange takes place through a mask opening, ions diffuse beneath the masked part generating a transition region (s), which extends beyond the opening (w) (see Figure 1).

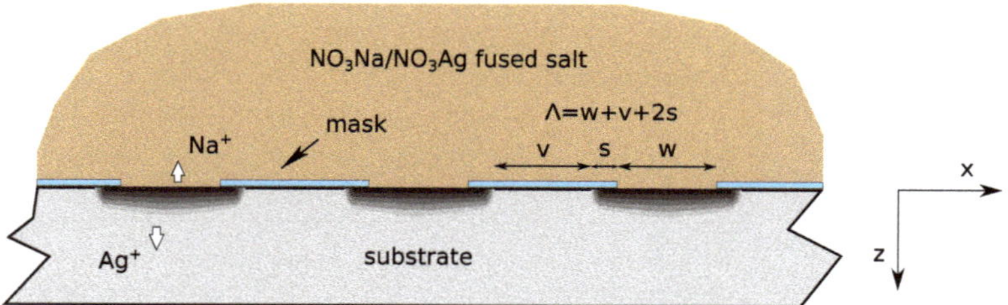

Figure 1. Scheme of the fabrication of an ion-exchanged grating. s and Λ indicate the transition region (or side diffusion) and the period of the grating, respectively.

Then, the phase profile of an ion-exchanged grating, $\Delta\varphi(x)$, would be more like a trapezoid. Moreover, the openings are usually similar in size to the transition region. This makes the ions spread deeper in the middle of the openings than close to their edges. As a result, the phase profile should be rounded in the region corresponding to the opening, just as predicted by the diffusion equation that governs the ion-exchange process [7] and confirmed experimentally [13]. In a first approximation, we can assume, for this region, a triangular phase profile and a linear profile for the transition region or side diffusion. Therefore, all the phase profile can be modeled by using straight lines giving place to the so-called polygonal model (see Figure 2).

In order to gain more insight into this shape of the phase profile, we plot, in Figure 3, typical profiles of in-depth refractive index changes of a substrate, $\Delta n(z)$, for three side positions: at the middle of the opening, at the edge of the mask, and at the middle of the transition region. As is known, this change of the refractive index is proportional to the concentration $C(x,z)$ of in-diffused ions, which can be calculated by solving a two-dimensional non-linear diffusion equation [7]. From $\Delta n(x,z)$, the phase profile is calculated by [14]:

$$\Delta\varphi(x) = \frac{2\pi}{\lambda} \int_0^\infty \Delta n(x,z)dz, \tag{1}$$

where λ is the wavelength. From this equation, it can be easily checked that the shape of the polygonal phase profile is compatible with the curves shown in Figure 3. On the other hand, the third dimension (y-axis) has little effect on the phase profile, and hence on the grating behavior, because the diffusion length of ions is very short compared to the length of this grating dimension. In other words, the ion-diffusion is not limited in the y-direction.

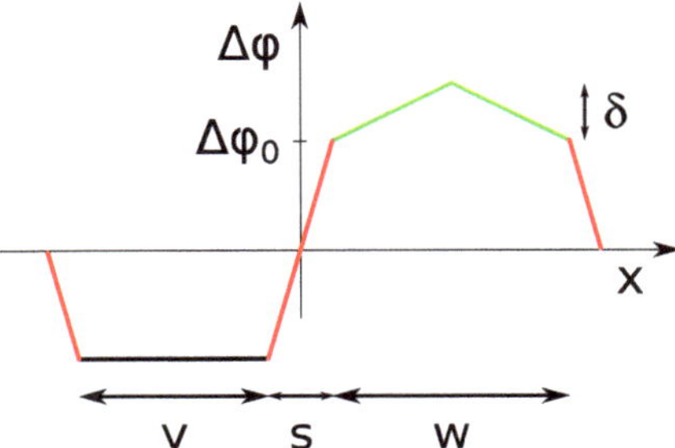

Figure 2. Modeling of the phase profile of an ion-exchanged grating by the polygonal model.

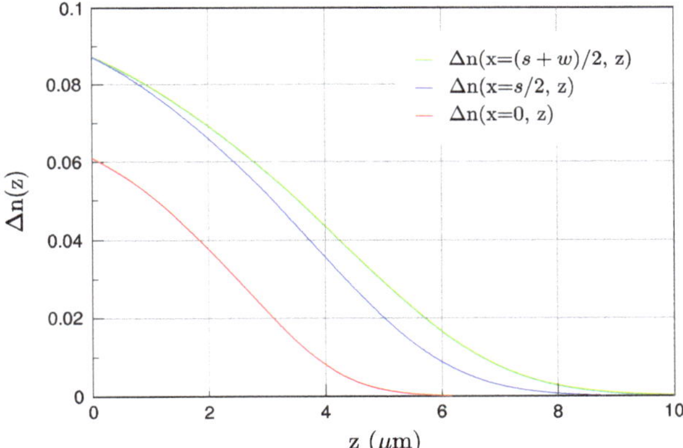

Figure 3. Typical in-depth refractive index changes after an ion-exchange process through an opening for three side positions: middle of the opening (green curve), the edge of the mask (blue line), and the middle of the transition region (red line).

According to Figure 2, the phase profile can be expressed as follows:

$$\Delta\varphi(x) = \begin{cases} -2\Delta\varphi_0\left(1 + v/s + x/s\right) & -(v+s) \le x \le -(v+s/2) \\ -\Delta\varphi_0 & -(v+s/2) \le x \le -s/2 \\ 2\Delta\varphi_0\, x/s & -s/2 \le x \le s/2 \\ \Delta\varphi_0 + \delta(-s/w + 2x/w) & s/2 \le x \le (w+s)/2 \\ \Delta\varphi_0 + \delta(2 + s/w - 2x/w) & (w+s)/2 \le x \le w+s/2 \\ 2\Delta\varphi_0\left(1 + w/s - x/s\right) & w+s/2 \le x \le w+s \end{cases} \tag{2}$$

where all the variables are defined in that figure. This model allows the characterization of the two-dimensional ion-exchange process, which takes place in the fabrication of the grating through the determination of the parameters of the model (s, δ, and $\Delta\varphi_0$). In Section 3, we will explain the procedure that we used. It is based on the comparison between the theoretical and experimental efficiencies of the gratings fabricated and will show the utility of this simple model.

For more complex periodic structures, the phase profile would be given by the following general polygonal function:

$$\Delta\varphi(x) = \{a_j + b_j x, \quad x_{j-1} \leq x \leq x_j, \quad j = 1, ..., M, \quad x_0 = -\frac{\Lambda}{2}, x_M = \frac{\Lambda}{2}\}, \tag{3}$$

where Λ is the period of the grating, M the number of segments, and a_j, b_j would be determined by the specific shape of the profile. The substitution of this expression in the equation for the transmission of the order m of a grating, in the thin-grating approximation,

$$t_m = \frac{1}{\Lambda} \int_{-\Lambda/2}^{\Lambda/2} \exp(i\Delta\varphi(x)) \exp(-im\frac{2\pi}{\Lambda}x)dx, \tag{4}$$

leads to a general expression for the diffraction coefficients of such a grating:

$$t_m = \sum_j \frac{\exp(ia_j)}{i(\Lambda b_j - 2\pi m)} \{\exp[i(b_j - \frac{2\pi m}{\Lambda})x_j] - \exp[i(b_j - \frac{2\pi m}{\Lambda})x_{j-1}]\}. \tag{5}$$

We must stress that in a standard selective ion-exchange process, the parameters a_j and b_j will be functions of ion-exchange time t. As shown later, a linear dependence between these parameters and $\sqrt{t}$ is found in a first approximation, that is,

$$a_j(t) = a_{j1} + a_{j2}\sqrt{t} \quad \text{and} \quad b_j(t) = b_{j1} + b_{j2}\sqrt{t}, \tag{6}$$

where a_{j1}, a_{j2}, b_{j1}, and b_{j2} would be obtained after characterization of the ion-exchange process. In general, different cases are possible: $a_{j1} = a_{j2} = 0$; $a_{j1} = 0$, $a_{j2} \neq 0$; $a_{j1} \neq 0$, $a_{j2} = 0$; and the same cases for b_{j1} and b_{j2}. This allows us to analyze the temporal evolution of diffraction coefficients, $t_m(t)$, so that an energy transfer is observed between diffraction orders m, giving rise to different optical transformations. As for the profile given by Equation (2), it corresponds to the particular case $M = 6$ with $b_2 = 0$, where the period Λ equals $w + v + 2s$. In this case, by introducing Equation (2) into Equation (4) and integrating, we obtain, after a long but straightforward calculation, the following expression for the diffraction coefficient of order m:

$$\begin{aligned}
t_m =& \frac{s}{\Lambda} \frac{\sin(\Delta\varphi_0 - \pi ms/\Lambda)}{\Delta\varphi_0 - \pi ms/\Lambda} + \frac{s}{\Lambda} \exp(i\frac{\pi m}{\Lambda}(v - w)) \frac{\sin(\Delta\varphi_0 + \pi ms/\Lambda - m\pi)}{\Delta\varphi_0 + \pi ms/\Lambda} \\
&- \exp(i\frac{\pi m}{\Lambda}(v - w)) \frac{\sin(\Delta\varphi_0 + \pi ms/\Lambda - m\pi)}{m\pi} + \frac{\sin(\Delta\varphi_0 - \pi ms/\Lambda)}{m\pi} \\
&+ \frac{w}{\Lambda} \exp(i\Delta\varphi_0) \exp(-i\frac{\pi m}{\Lambda}(s + w)) \left(\exp(i(\frac{\pi mw}{2\Lambda} + \frac{\delta}{2})) \frac{\sin((\pi mw/\Lambda - \delta)/2)}{\pi mw/\Lambda - \delta}\right) \\
&+ \frac{w}{\Lambda} \exp(i\Delta\varphi_0) \exp(-i\frac{\pi m}{\Lambda}(s + w)) \left(\exp(-i(\frac{\pi mw}{2\Lambda} - \frac{\delta}{2})) \frac{\sin((\pi mw/\Lambda + \delta)/2)}{\pi mw/\Lambda + \delta}\right) \\
&- \frac{w}{\Lambda} \exp(i\Delta\varphi_0) \exp(-i\frac{\pi m}{\Lambda}(s + w)) \frac{\sin(\pi mw/\Lambda)}{\pi mw/\Lambda}.
\end{aligned} \tag{7}$$

The first two lines are the transmission for $\delta = 0$ and the last ones the correction to take into account the triangle defined in the interval $[s/2, w + s/2]$. Note that the diffraction efficiency of the order m can be easily calculated as:

$$\eta_m = |t_m|^2. \tag{8}$$

3. Characterization of the Two Dimensional Ion-Exchange Process by the Polygonal Model

The polygonal model presented in the previous section enables the characterization of the ion-exchange process, which takes place during the grating fabrication. From this characterization, specific grating elements can be designed and fabricated. The detailed procedure is described here, and it consists of three steps.

First, a raw characterization of the one-dimensional ion-exchange process is made through the fabrication of some planar waveguides in suitable substrates. In our case, we fabricated these waveguides by using thermal ion-exchange in soda-lime glasses and a mixture 5% mole $AgNO_3/NaNO_3$, at $340\,^{\circ}C$, as the ion source. Next, the ion-exchange modification produced by this salt in this particular glass is analyzed by the inverse Wentzel–Kramers–Brillouin (IWKB) method. As is known, the IWKB method facilitates the determination of the in-depth refractive index profile resulting from the ion-exchange process. From this refractive index profile, the phase step and the penetration depth of the ions, for a given ion-exchange time, can be calculated [14]. In particular, we needed an ion-exchange time t = 13 min to get an in-depth penetration of about 7.5 μm and a phase step around π for a wavelength λ_0 = 632.8 nm.

Second, the geometric characteristics of the gratings are chosen. In particular, their period is usually selected from diffraction needs. We chose a period Λ = 30 μm and an opening width w = 10 μm. The size of the opening was chosen to obtain a grating with the two sides of its phase profile similar in size, that is a quasi-antisymmetric phase profile. Note that most interesting phase profiles are of this kind. This opening width leads to $v \approx$ 10 μm and $s \approx$ 5 μm for a phase step π, because the side diffusion of the ions is somewhat smaller than the in-depth diffusion. Next, the grating must be patterned on the substrate. We used for it a two-step photolithographic process. In the first step, a binary grating pattern with the aforementioned dimensions was plotted on a sheet of heavy paper and then reduced $-1/13.5$ by using a lens Schneider XENON Sapphire 4.5/95. This led to a chromium master on a glass substrate that was then reduced by a Carl Zeiss S-planar lens working at magnification $M = -1/5$ and $f/\# = 1.3$. The final result was an aluminum binary grating on the soda-lime glass chosen for the ion-exchange.

The third step is the characterization of the two-dimensional ion-exchange process that takes place during the grating fabrication. For that, as we will see, it is enough to use three gratings with approximate phase steps $\pi/2$, π, and $3\pi/2$. These phase steps cover the range of phases $[0–2\pi]$, which almost any interesting phase grating element will have. Then, from the characterization realized in the first step, we calculated the ion-exchange times (3.5 min, 14.5 min, and 32 min) needed to achieve, approximately, these phase steps. Next, after the fabrication of these three gratings, their diffraction efficiencies (η_m^{exp}) were measured and compared to those provided by the polygonal model (Equations (7) and (8)). This is done by calculating the RMS of the differences between measured and calculated efficiencies, that is:

$$RMS(w,s,\delta,\Delta\varphi_0) = \sqrt{\sum_{m=-(N-1)/2}^{(N-1)/2} (\eta_m^{exp} - \eta_m^{Equation(7)})^2/N}, \qquad (9)$$

where N is the number of the diffraction orders measured. The minimization of this RMS permitted us to fit the polygonal model to the exact phase profile of each grating. Note that in addition to the intrinsic parameters of the model, we also considered the width of the opening as a free parameter. This was done to take into account the fabrication tolerances and/or to allow the approximate model to fit better to the real profile.

We did the minimization task by using the function fmincon of the software Octave [15]. This minimization was made in two passes. In the first one, we let the four parameters vary freely, except for w, which we constrained to values close to the real opening width (w = 10 $\pm$1.5 μm). After this, several findings can be highlighted. First, all three minima of the RMSs were under 0.05%, which confirms the validity of the model.

Second, w varied approximately between 8.5 and 10 μm. Third, a good linear relationship was observed between s and $\Delta\varphi_0$ and $\sqrt{t}$. This behavior was expected because the penetration depth in ion-exchange processes, as well as the phase steps that ion-exchange yields depend linearly on the square root of time [14]. Finally, the dependence between δ and $\sqrt{t}$ was not clearly linear. However, it is reasonable to guess a linear dependence between these two magnitudes because δ is a part of the maximum phase step $(2\Delta\varphi_0 + \delta)$, and this should depend linearly on $\sqrt{t}$, at least for short times. Moreover, we checked that the minima of the RMSs were not too sensitive to changes in δ. With this in mind, we did a second minimization pass where the values of δ were constrained to the best linear fit obtained in the first pass $\pm 10\%$, and w was fixed to the average of the three values resulting from this first fit, that is 9.1 μm. Note that this value of the opening width is remarkably different from the real value (10 μm) of the fabricated mask. This is a consequence of the coarse approximation of the phase profile that we obtained with the polygonal model. After the second minimization pass, the three minima of the RMSs were under 0.08%, and we obtained the linear dependencies shown in Figure 4.

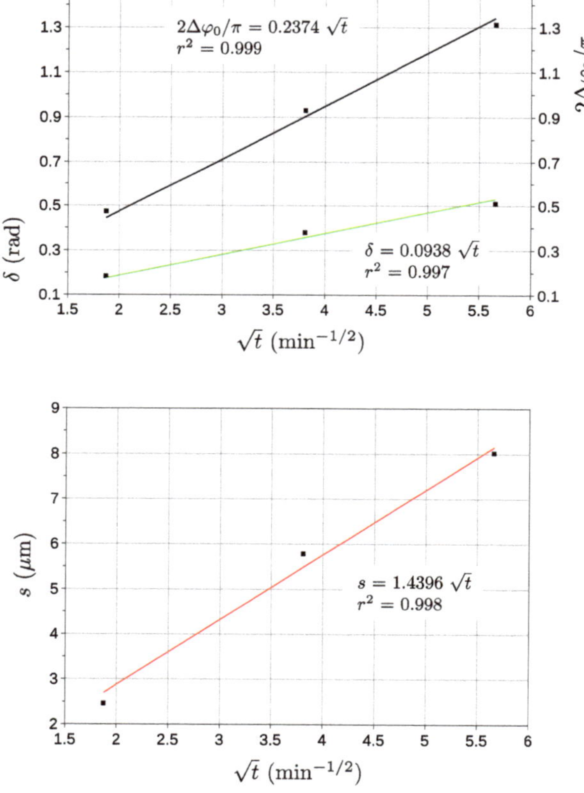

Figure 4. Data points of the polygonal model that best predicted the experimental efficiencies of the three gratings fabricated to characterize the ion-exchange process; the lines of the best fit as functions of the square root of the ion-exchange time are also plotted. The parameter w was fixed to 9.1 μm.

This figure confirms the linear dependencies already observed in the first pass. However, a closer examination reveals a slight departure from the linearity of $\Delta\varphi_0$ and δ. This is a consequence of the saturation of the in-depth diffusion of ions for long times. Anyway, as we will see in the next section, it did not have any apparent impact on the design and fabrication of, at least, the two elements that we made. On the other hand, we checked that the inclusion of more than three gratings in the above characterization did not change the quality of the fits or the predictions. Now, if we insert in Equation (7) the linear equations shown in this figure, only the ion-exchange time remains as the independent variable. Therefore, we can represent the efficiencies of the diffraction orders (Equation (8)) as functions of $\sqrt{t}$ (see Figure 5). This figure includes the efficiencies of the three gratings used in the characterization process, as well as those of the two grating elements that we designed and fabricated using such a characterization, as we will see in the next section.

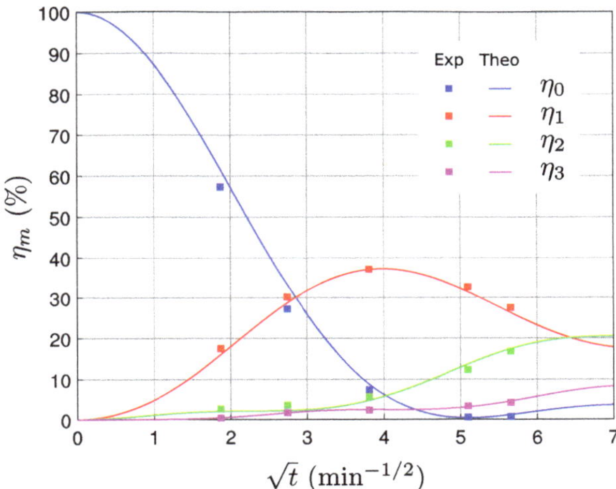

Figure 5. Plot of the efficiencies of the orders 0 (blue line), ±1 (red line), ±2 (green line), and ±3 (magenta line) with the square root of the ion-exchange time according to Equations (7) and (8) and Figure 4. The experimental efficiencies of the five gratings used in this work are also plotted.

4. Design and Fabrication of a Grating with Suppressed Zero Order and a Three Splitter

Now, we show the practical utility of the model introduced in Section 2 by designing and fabricating two diffractive elements: a grating that suppresses its zero order and a three splitter. The procedure that we used for this was the same in both cases. First, we calculated the ion-exchange time necessary to obtain each element through Equations (7) and (8), as well as the characterization presented in Section 3 (Figures 4 and 5). Next, we fabricated such an element following the same fabrication process explained in Section 3, but using the previous calculated time. Finally, we measured the diffraction efficiencies for $\lambda_0 = 632.8$ nm to check the quality of the element. Along these measurements, we also checked that these gratings had negligible absorption losses at this wavelength and no sensitivity to polarization of the beam. Moreover, we quantified their noise by measuring the stray light between 0–1 orders and 1–2 orders. We obtained a signal-to-noise ratio higher than 35 dB. The experimental setup to measure the diffraction efficiency of these grating elements is shown in Figure 6, and a close up photo of one of the fabricated gratings can be seen in Figure 7.

Of course, the design of these elements was restricted to the parameters of the gratings used to characterize the ion-exchange process, that is a period $\Lambda = 30$ μm and an opening width $w = 10$ μm. However, other periods and openings could be used provided that a previous characterization is done. In particular, by using this polygonal model, the period and the opening of the gratings could be optimized for a specific element on the basis of, for instance, a large fabrication tolerance or another merit function. On the other hand, this model provides the minimum period that can be achieved using this technique of grating fabrication. From Figures 1 and 2, it is clear that the limit is found for $v = 0$, that is $\Lambda_{min} = w + 2s$. The particular value will depend on s, which will be determined by the particular element to be fabricated. For example, for a grating with suppressed zero order (see Section 4.1), $\Lambda_{min} \approx 15$ µm, and for a three splitter (see Section 4.2), $\Lambda_{min} \approx 9$ µm. In both cases, an opening width of $w \approx 1$ µm is assumed.

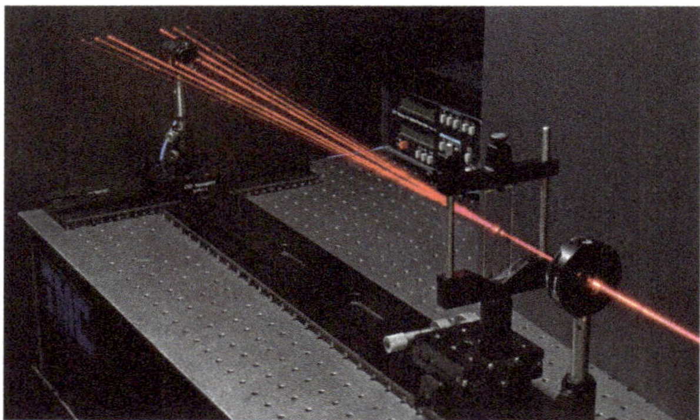

Figure 6. Long exposure photography of the experimental setup. On the right side, a polarizer, before one of the fabricated gratings, was placed to modify the intensity of the beam. On the left side, you can see the detector head, with a diaphragm on its front. The stabilized He-Ne laser used in the measurements is not shown in the picture.

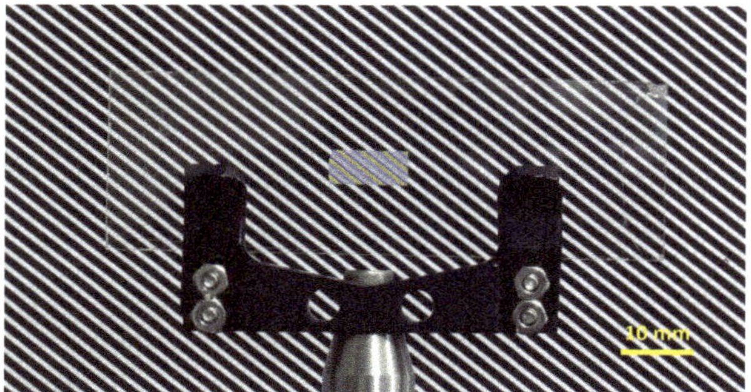

Figure 7. Close-up photography of one of the fabricated gratings (in the middle of the picture). Behind, you can see a periodic object to better observe the dimensions of the grating with respect to the glass substrate. The grating grooves are arranged horizontally.

4.1. Design and Fabrication of a Grating with Suppressed Zero-Order

The cancellation of the zero order implies $\eta_0 = 0$. According to the blue curve (efficiency of the zero order) of Figure 5, this condition is fulfilled for $t \approx 26$ min. Therefore, we fabricated a new grating using this ion-exchange time. After fabrication, we measured the diffraction efficiencies and obtained the results shown in Table 1.

Table 1. Experimental and predicted diffraction efficiencies for λ_0 = 632.8 nm of the grating designed to suppress its zero order.

m	$|t_m^2|_{exp}$	$|t_m^2|_{Equation(7)}$
0	0.5%	0.35%
±1	32.55%	31.4%
±2	12.2 %	13.6%
±3	3.25 %	3.0%
±4	1.3%	1.3%
±5	0.2%	0.3%
±6	0.05%	0.05%

The result was very good. The efficiency of the zero order did not reach the theoretical minimum, but was very close, and the efficiency of the first orders (±1) was higher than expected. This closeness between the predicted and obtained values was not surprising because of the low sensitivity of the diffraction efficiency η_0, at its minimum, with the ion-exchange time (see Figure 5). Furthermore, the efficiency obtained at the first orders was slightly higher than the value (27%) that a sinusoidal phase grating provides. If higher efficiencies were necessary, we could stack a second grating with half the period to obtain a non-symmetrical phase profile. A proper design of these two phase elements would result in an improved efficiency for the +1 order at the expense of the −1 order. On the other hand, the small discrepancy between the experimental and theoretical values of the efficiencies can be attributed to the polygonal model, which does not perfectly reproduce the real phase profile. Anyway, the small values η_0 that we obtained validate this ion-exchange technique to fabricate this kind of gratings and the polygonal model to represent its phase profile.

4.2. Design and Fabrication of a Three Splitter

The procedure that we used to fabricate a three splitter was the same as described in the previous paragraphs, but now the condition $\eta_0 = \eta_1$ or $\eta_0 = \eta_{-1}$ must be fulfilled. Therefore, from Figure 5, we must calculate the intersection between the blue curve (efficiency of the zero order) and the red curve (efficiency of the first order). This intersection is at $t \approx 8.1$ min; however, we finally rounded this time to $t = 8$ min because of the experimental uncertainties of the fabrication process. Indeed, for such a short time, the transient regimes are relevant, and we observed that they gave rise to uncertainties of about 3% in the ion-exchange time. After fabrication, we measured the diffraction efficiencies of the first three orders. They were $\eta_0 = 25.5\%$ and $\eta_{\pm1} = 31.0\%$. It is clear from these values that our goal was not achieved. Three reasons can be given to explain such a discrepancy. First, is the inaccuracy of the polygonal model; second is the error in the opening width resulting from the photolithographic process; and third is the high sensitivity, in this region, of the efficiencies of the first two orders with the duration of the ion-exchange process, which along with the aforementioned uncertainties made the prediction very difficult. Both the second and third reasons can be minimized by using a thorough lithography and lower temperatures for the ion-exchange process, respectively. Despite the differences obtained between the efficiencies, note that the model gave a result close to the desired one, as well as clues to find a better solution. Indeed, according to Figure 5, the time was overestimated because the efficiencies obtained were on the right side of the intersection, at $t \approx 9$ min. Therefore, we considered a new ion-exchange time $t = 7.5$ min, expecting to achieve a better result. The diffraction efficiencies of this last grating are shown in Table 2. The values were not yet optimal, but there was no point in doing more trails due to the reasons given before. Anyway, this result can be considered good: the difference between the efficiencies of the first three orders was small (3%), and their sum (87.3%) was near the theoretical maximum (92.5%). This last fact was a consequence of the similarity between the phase profile resulting from this fabrication process and the profile necessary to have an optimal three splitter [16]. It is true that the difference between the experimental and predicted efficiencies was a little high (about 6% for the zero order). However, this polygonal model was not conceived of to predict exactly the shape of the phase profile and hence the efficiencies of the gratings. Our goal was to provide a simple model to design optical elements made of ion-exchanged gratings.

Table 2. Experimental and predicted diffraction efficiencies for $\lambda_0 = 632.8$ nm of the grating designed to be a three splitter.

m	$|t_m^2|_{exp}$	$|t_m^2|_{Equation(7)}$
0	27.1%	33.3%
±1	30.1%	28.65%
±2	3.5%	2.15%
±3	1.6%	1.55%
±4	0.7%	0.5%
±5	0.3%	0.25%
±6	0.1%	0.1%

Finally, we show in Figure 8 the image of a red pencil (left) through the grating with suppressed zero order (middle) and the three splitter (right). Of course, this result agreed with the data given in the previous tables. With regard to the grating that suppressed its zero order, only the blue component of the pencil remained at this order because the red component was completely removed. Moreover, the ±1 orders were the most relevant ones. As for the three splitter, the three first orders were approximately equal in brightness, and the rest of the orders almost vanished.

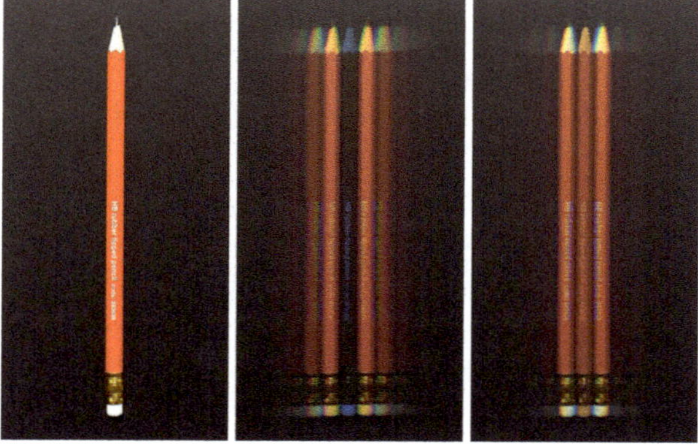

Figure 8. Images of a red pencil (**left**) through the grating with suppressed zero order (**middle**) and the three splitter (**right**).

5. Conclusions

We present a simple polygonal model to describe the phase profile of ion-exchanged gratings. Its simplicity makes it easy to calculate the transmission function of gratings with such a phase profile. From this transmission function and the experimental efficiencies of three gratings with approximate phase steps $\pi/2$, π, and $3\pi/2$, we characterized the ion-exchange process through simple functions for the intrinsic parameters of the model. In order to show the potential of this model, we designed and fabricated a diffraction grating with suppressed zero order and a three splitter. The cancellation of the zero order was very good: only 0.5% of the power remains at this order after diffraction. As for the three splitter, its phase profile is close to the optimal one, and the difference between the efficiencies of its first orders is 3%. Moreover, this value can be further improved by lowering the temperature of the ion-exchange process. Finally, this polygonal model could be useful to design more complex diffractive elements by increasing the number of segments. It could also be generalized to the two-dimensional case.

Author Contributions: Conceptualization, X.P.-B. and J.L.; methodology, C.M.-O., X.P.-B., H.G.-N., and J.L.; software, C.M.-O. and H.G.-N.; validation, C.M.-O. and H.G.-N.; visualization, C.M.-O. and H.G.-N.; investigation, C.M.-O., X.P.-B., H.G.-N., and J.L.; writing, original draft preparation, C.M.-O.; writing, review and editing, C.M.-O., X.P.-B., H.G.-N., and J.L. All authors read and agreed to the published version of the manuscript.

Funding: This research was funded by Xunta de Galicia, Consellería de Educación, Universidades e FP, Grant GRC number ED431C2018/11, and Ministerio de Economía, Industria y Competitividad, Gobierno de España, Grant Number AYA2016-78773-C2-2-P.

Institutional Review Board Statement: Not applicable.

Informed Consent Statement: Not applicable.

Data Availability Statement: Data is contained within the article.

Conflicts of Interest: The authors declare no conflict of interest.

References

1. Hariharan, P. *Optical Holography*; Cambridge University Press: Cambridge, UK, 1996.
2. Gauglitz, G.; Vo-Dinh, T. *Handbook of Spectroscopy*; WILEY-VCH Verlag GmbH & Co. KGaA: Weinheim, Germany, 2003.
3. Cagigal, M.P.; Valle, P.J. Wavefront sensing using diffractive elements. *Opt. Lett.* **2012**, *37*, 3813–3815. [CrossRef] [PubMed]
4. Linares, J.; Montero, C.; Moreno, V.; Nistal, M.C.; Prieto, X.; Salgueiro, J.R.; Sotelo, D. Glass processing by ion exchange to fabricate integrated optical planar components: Applications. *Proc. SPIE Int. Soc. Opt. Eng.* **2000**, *3936*, 227–238. [CrossRef]
5. Montero-Orille, C.; Moreno, V.; Prieto-Blanco, X.; Mateo, E.F.; Ip, E.; Crespo, J.; Liñares, J. Ion-exchanged glass binary phase plates for mode-division multiplexing. *Appl. Opt.* **2013**, *52*, 2332–2339. [CrossRef] [PubMed]
6. Salmio, R.P.; Saarinen, J. Graded-Index Diffractive Elements by Thermal Ion Exchange in Glass. *Appl. Phys. Lett.* **1995**, *66*, 917–919. [CrossRef]
7. Salmio, R.P.; Saarinen, J.; Turunen, J.; Tervonen, A. Graded-Index Diffractive Structures Fabricated by Thermal Ion Exchange. *Appl. Opt.* **1997**, *36*, 2048–2057. [CrossRef] [PubMed]
8. Chen, G.L.; Lin, C.Y.; Kuo, M.K.; Chang, C.C. Numerical suppression of zero order image in digital holography. *Opt. Express* **2007**, *15*, 8851–8856. [CrossRef] [PubMed]
9. Improso, W.D.G.D.; Tapang, G.A.; Saloma, C.A. *Suppression of Zeroth-Order Diffraction in Phase-Only Spatial Light Modulator*; Springer Series in Optical Sciences; Springer: Cham, Switzerland, 2019; Volume 222.
10. Mihailov, S.J.; Smelser, C.W.; Lu, P.; Walker, R.B.; Grobnic, D.; Ding, H.; Henderson, G.; Unruh, J. Fiber Bragg gratings made with a phase mask and 800-nm femtosecond radiation. *Opt. Lett.* **2003**, *28*, 995–997. [CrossRef] [PubMed]
11. Pöyhönen, P.; Honkanen, S.; Tervonen, A.; Tahkokorpi, M.; Albert, J. Planar 1/8 Splitter in Glass by Photoresist Masked Silver Film Ion Exchange. *Electron. Lett.* **1991**, *27*, 1319–1320. [CrossRef]
12. Marco, D.; Sánchez-López, M.M.; Cofré, A.; Vargas, A.; Moreno, I. Optimal triplicator design applied to a geometric phase vortex grating. *Opt. Express* **2019**, *27*, 14472–14486. [CrossRef]
13. Singer, W.; Dobler, B.; Schreiber, H.; Brenner, K.H.; Messerschmidt, B. Refractive-index measurement of gradient-index microlenses by diffraction tomography. *Appl. Opt.* **1996**, *35*, 2167–2171. [CrossRef]
14. Prieto-Blanco, X.; Montero-Orille, C.; Moreno, V.; Mateo, E.F.; Liñares, J. Chromatic characterization of ion-exchanged glass binary phase plates for mode-division multiplexing. *Appl. Opt.* **2015**, *54*, 3308–3314. [CrossRef] [PubMed]
15. Eaton, J.W.; Bateman, D.; Hauberg, S.; Wehbring, R. GNU Octave, Version 4.4.0 Manual: A High-Level Interactive Language for Numerical Computations. Network Theory Limited. 2018. Available online: https://octave.org/doc/v4.4.1/ (accessed on 2 June 2020).
16. Gori, F.; Santarsiero, M.; Vicalvi, S.; Borghi, R.; Cincotti, G.; di Fabrizio, E.; Gentili, M. Analytical derivation of the optimum triplicator. *Opt. Commun.* **1998**, *157*, 13–16. [CrossRef]

applied
sciences

Article

Governing Functionality of Silver Ion-Exchanged Photo-Thermo-Refractive Glass Matrix by Small Additives

Yevgeniy Sgibnev *, Nikolay Nikonorov and Alexander Ignatiev

Faculty of Photonics, ITMO University, 199034 St Petersburg, Russia; nikonorov@oi.ifmo.ru (N.N.); ignatiev@oi.ifmo.ru (A.I.)
* Correspondence: sgibnevem@gmail.com

Abstract: In this study, the influence of small additives on the spectral and optical properties of Na^+–Ag^+ ion-exchanged silicate glass is presented. Polyvalent ions, for example, cerium and antimony, are shown to reduce silver ions to atomic state and promote the growth of photoluminescent silver molecular clusters and plasmonic silver nanoparticles. Na^+–Ag^+ ion-exchanged and heat-treated glasses doped with halogen ions, such as chlorine or bromine, exhibit formation of photo- and thermochromic AgCl or AgBr nanocrystals. Growth of a silver nanoisland film on the glass surface was observed in the case of undoped sample. The presented results highlight the vital role of small additives to control properties of the silver nanostructures in Na^+–Ag^+ ion-exchanged glasses. Possible applications of Na^+–Ag^+ ion-exchanged glass ceramics include but are not limited to biochemical sensors based on surface-enhanced Raman scattering phenomena, temperature and overheating sensors, white light-emitting diodes, and spectral converters.

Keywords: ion exchange; silver; photo-thermo-refractive glass; silver nanoparticles; silver molecular clusters; silver bromide nanocrystals

check for **updates**

Citation: Sgibnev, Y.; Nikonorov, N.; Ignatiev, A. Governing Functionality of Silver Ion-Exchanged Photo-Thermo-Refractive Glass Matrix by Small Additives. *Appl. Sci.* **2021**, *11*, 3891. https://doi.org/10.3390/app11093891

Academic Editor: Jesús Liñares Beiras

Received: 14 March 2021
Accepted: 20 April 2021
Published: 25 April 2021

Publisher's Note: MDPI stays neutral with regard to jurisdictional claims in published maps and institutional affiliations.

1. Introduction

Over time, glasses have occupied a particular place among various optical materials due to their wide range of transparency, easily tuned properties, and workability. Recently, research has focused more on the development of new composite glassy materials, e.g., including materials doped with nanoparticles and/or nanocrystals. Today, functional glasses that actively interact with light are required. An example of functional optical material is photo-thermo-refractive (PTR) glass based on a sodium–zinc–aluminosilicate matrix containing halogens (fluorine and bromine) and doped with silver, antimony, and cerium oxides [1,2]. The refractive index of PTR glass is changed after exposure to UV light and the subsequent thermal treatment in the vicinity of the glass transition temperature. The latter step leads to the formation of silver nanoparticles (SNPs) playing the role of nucleation centers for NaF nanocrystals in the glass host, causing variations in the refractive index. Mechanisms of photo-thermo-induced crystallization in PTR glass were studied in detail and reported in [3,4]. Currently, phase holograms recorded in PTR glass are used as line-narrowing and stabilizing filters, spectral and spatial filters, compressors for femto- and picoseconds laser beams, spectral beam combiners, high-power beam splitters, etc. Although PTR glass was developed as a photosensitive medium for recording volume Bragg gratings, recent studies showed that it is a polyfunctional material that can combine the properties of holographic, laser, and photoetchable media [1,2,5–8]. In this work, we demonstrate that the functionality of PTR glass matrix can be substantially extended via the formation of silver nanostructures inside the glass or on its surface.

Silver can be introduced in a silicate glass matrix by adding silver-containing chemicals into the glass batch, via ion implantation or via ion exchange (IE). Due to the low solubility of silver in a silicate matrix, the first approach allows embedding up to 0.2 mol.% Ag_2O. Ion implantation requires special sophisticated equipment, and only very thin layers can be modified with the method, whereby the typical thickness of the silver-enriched region is below 2 μm [9,10]. The IE technique, on the other hand, is characterized by simplicity, flexibility, and applicability for mass production. For example, Na^+–K^+ IE is used for glass strengthening on an industrial scale for producing cover glass for smartphones [11]. The method is based on the substitution of alkali (Na^+, K^+, Rb^+, Cs^+) or transition (Ag^+, Cu^+, Tl^+) metal ions from glass by another one from the salt melt, metal film, or another source [12–14]. A comparison of ion implantation and ion exchange technologies for modification of glass is shown in Table 1.

Table 1. Comparison of the ion implantation and ion exchange technologies.

Parameter	Ion Exchange	Ion Implantation
Simplicity/complication	Very simple and flexible technology	Requires sophisticated equipment
Applications of modified glass	Optical waveguides; glass strengthening; nanostructuring	Optical waveguides; glass hardening; nanostructuring
Variety of metal nanostructures	Limited to copper and silver nanostructures due to possibility of ions to effectively exchange with alkali ions from a glass substrate	Depends on ion source: Ag, Cu, Ni, Co, Ti, Cr, and Zn nanoparticles were obtained and studied in glasses
Requirements for glass substrate	High chemical durability to salt melt; limited to alkali-containing glasses	No requirements
Thickness of the glass modified region	The range is very wide and depends on particular application, typically in the range from few to tens of microns	Typically, below 2 μm
Temperature during modification	Should be at least above melting temperature of salt melt; in most cases, temperature is in the range 250–450 °C.	Low-temperature processing
Reproducibility	Excellent	Excellent
Allows overcoming solubility of dopant	Yes	Yes
High-scale production capability	Already used for mass production	Already used for mass production
Allows overcoming solubility of dopant in glass matrix	Yes	Yes

It must be noted that most research papers devoted to the study of silver nanostructures formed in ion-exchanged glasses focused on the influence of the ion exchange (concentration of silver in salt melt, temperature, and duration) and subsequent heat treatment (temperature and duration) parameters, while glass composition and influence of the glass additives have not received much attention. As it is shown further, these uncharacterized factors play a key role in determining the properties of silver nanostructures and, consequently, the functionality of the glass doped with them. This research deals with the influence of typical additives of PTR glass on the spectral and optical properties of Na^+–Ag^+ ion-exchanged layers. However, as PTR glass is based on the well-known and widely used sodium–aluminosilicate matrix, our results can be extended to other silicate glasses.

2. Materials and Methods

Glass blocks of samples based on the $14Na_2O$–$3Al_2O_3$–$5ZnO$–$71.5SiO_2$–$6.5F$ (mol.%) composition were synthesized in a furnace at 1500 °C in air atmosphere using a platinum crucible and mechanical stirrer. The glass transition temperature of the samples was determined using an STA 449F1 Jupiter (Netzsch, Germany) differential scanning calorimeter at a heating rate of 10 K/min. For the optical study, planar polished samples with a thickness of 0.9–1.0 mm were prepared. Na^+–Ag^+ IE was conducted by immersing glass samples in a bath with a melt of silver and sodium nitrates (5 mol.% $AgNO_3$/95 mol.% $NaNO_3$) at 320 °C. Absorbance spectra were recorded with a double-beam spectrophotometer Lambda 650 (Perkin-Elmer). The emission spectra were measured inside an integrated sphere using a photonic multichannel analyzer (PMA-12, Hamamatsu, Japan) at room temperature. XRD characterization was performed using a Rigaku Ultima IV X-ray diffractometer.

3. Results and Discussion

3.1. Spontaneous Growth of Silver Nanoparticles on the Surface of Ion-Exchanged Photo-Thermo-Refractive Glass Heat-Treated in Air Atmosphere

Plasmonic metal nanoparticles have attracted significant interest in photonics due to their unique optical properties. Potential applications of metal nanoparticles include but are not limited to biosensing [15], surface-enhanced Raman spectroscopy (SERS) [16], solar cells [17], targeted drug delivery [18], saturable absorbers [19], and photocatalysis [20]. SERS is one of the most widely studied applications of the metal nanoparticles. The phenomenon was firstly observed from pyridine molecules adsorbed on roughened silver electrodes by Fleischmann et al. in 1974 [21]. Despite SERS mechanisms still being under discussion, it has become a powerful and ultrasensitive tool due to the possibility of enhancing Raman scattering magnitude by 10–14 orders. This allows one to detect even a single molecule [22,23].

Roughened metal surfaces can be obtained in different ways, including thermal, electron beam, chemical, and plasma deposition methods. Recently, it was shown that a SERS-active silver nanoisland film (SNIF) can be easily formed on the surface of Na^+–Ag^+ ion-exchanged glasses. Two different approaches were developed to grow SERS-active SNIF on a glass surface (Figure 1). The first is based on the growth of SNPs in the ion-exchanged glass layers with subsequent chemical etching to obtain SNPs directly on the glass surface [24,25]. The concentration of NPs on glass surface can be controlled by silver content in a salt melt during the IE process and by the thickness of the etched glass layer. The second approach involves thermal treatment of Na^+–Ag^+ ion-exchanged glass in a reducing atmosphere, for example, hydrogen atmosphere or water vapor [26,27]. Growth of SNPs in the subsurface glass region can be suppressed by low temperature of the treatment (at least 100 °C below the glass transition temperature). Moreover, the poling technique can be applied to Na^+–Ag^+ ion-exchanged glasses to obtain single NPs on the glass surface [28,29]. It should be noted that the considered approaches show similar performance as SERS-active substrates.

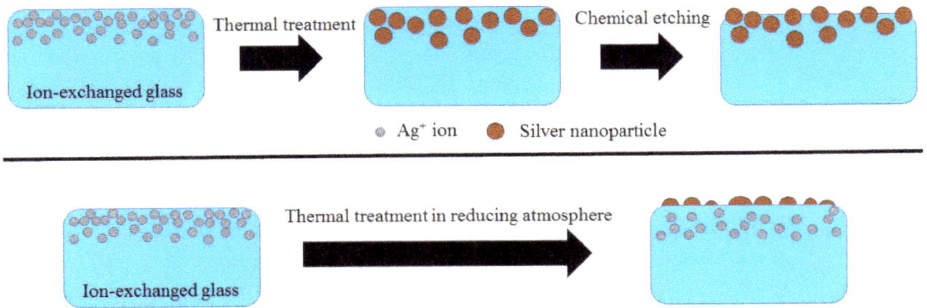

Figure 1. Schemes of two processes of formation of SNIF at the surface of Na^+–Ag^+ ion-exchanged glass.

Our preliminary research revealed the spontaneous growth of SNIF at the surface of Na^+–Ag^+ ion-exchanged undoped matrix-based PTR glasses. To our best knowledge, this is the first observation of growth of SNIF at the surface of ion-exchanged glass after heat treatment in air. Figure 2 shows photos of BK-7 optical glass and PTR matrix-based glass with no dopants (hereafter denoted as PTR-M; see Table 2) after the Na^+–Ag^+ IE and heat treatment at a temperature of 50 °C above corresponding T_g values. BK-7 exhibited the yellowish coloration typical for most silver ion-exchanged and heat-treated glasses, which results from absorption of the SNPs formed in silver-enriched glass layers [30,31]. PTR-M glass, on the contrary, possessed no coloration, due to the absence of electrons for reducing silver ions inside the glass [32]. Moreover, a gray semitransparent film was observed on the PTR-M glass surface.

Figure 2. Photo of the PTR-M glass sample (**left**) after Na^+–Ag^+ IE and heat treatment at 500 °C for 3 h (the resulting surface film was partially removed) and of BK-7 glass ion-exchanged and heat-treated at 600 °C for 3 h (**right**).

Table 2. Batch concentration of additives in the studied PTR glass samples.

Glass	Sb_2O_5, mol.%	Br, mol.%
PTR-M	0	0
PTR-Br	0	1.4
PTR-Sb2	0.002	0
PTR-Sb4	0.004	0
PTR-Sb10	0.01	0

For further detailed research, PTR-M glass with no dopants was studied. Silver ions were introduced by immersing the PTR glass samples into the bath containing a mixture of $5AgNO_3/95NaNO_3$ (mol.%) at 320 °C for 15 min. Ion-exchanged PTR glass samples were then heat-treated in air at temperatures in the range 350–500 °C, with a treatment duration of up to 7 h. An Na^+–Ag^+ ion-exchanged glass layer was mechanically removed from one side of the samples prior the heat treatment. Glass samples were placed in a muffle furnace with the silver-containing side facing upward.

Figure 3 shows absorbance spectra of the ion-exchanged PTR-M glass samples heat-treated in different conditions. The absorption band peak at 445 nm, related to the growth of SNIF on the ion-exchanged PTR-M glass surface, was observed already after the heat treatment for 15 min at 350 °C (Figure 3a). The increase in absorption amplitude, red shift of the absorption peak, and emergence of the second band around 355 nm occurred with a longer heat treatment duration. An analogous dependence of the absorbance spectra on the heat treatment time was obtained for the samples heat-treated at 400 and 450 °C (Figure 3b,c). Similar changes in the absorbance spectra of SNPs dispersed in water solution were observed as a function of increasing size from 46 to 287 nm by Kinnan and Chumanov [33]. The authors assigned the observed changes to quadrupolar, octupolar, and hexadecapolar plasmon modes in SNPs with a size of 128–287 nm. As heat treatment temperature governs the kinetics of redox reactions and diffusion process of SNPs growth, it also influences the mean size and size distribution of SNPs. The heat treatment at 500 °C, i.e., above the glass transition temperature (464 °C for PTR-M glass), resulted in almost structureless absorbance spectra due to very broad and overlapped bands.

To confirm that neither silver molecular clusters (SMCs) nor SNPs were formed in the ion-exchanged PTR-M glass during the heat treatment, the following experiment was conducted: a sample of the Na^+–Ag^+ ion-exchanged PTR-M glass was repeatedly heat-treated at 500 °C. The SNIF formed on the PTR glass surface was removed mechanically by polishing the sample for 3–5 s, and absorption spectra were recorded after each cycle. The absorbance spectra obtained during this experiment are shown in Figure 4a. Na^+–Ag^+ IE in PTR glass is known to induce a long-wavelength shift of the UV absorption edge [34]. The reason for the shift is strong absorption of the introduced silver ions that are characterized by the absorption envelope peak at 225 nm [35]. Thus, in contrast to the IE effect on absorbance spectra, a short-wavelength shift of the UV absorption edge was observed in the heat-treated samples. The spectra of the ion-exchanged and heat-treated PTR-M sample after removing SNIF showed no absorbance in the range above 300 nm, where the absorption bands of SMCs and SNPs in PTR glass are located [36]. This result also proved that all the features observed in the spectra presented in Figure 3 are related to SNIF formed on the PTR-M glass surface. Location of the UV absorption edge of the Na^+–Ag^+ ion-exchanged PTR-M glass is determined by the absorption of Ag^+ ions [37]. Thus, it is possible to roughly estimate the integral silver concentration in the heat-treated samples compared to nontreated ones using the Beer–Lambert law. The differential absorbance spectra obtained by subtracting the spectrum of the as-synthesized sample from the others are shown on Figure 4b. The differential absorbance decreased more than eightfold at 255 nm, from ~2.6 to ~0.3 after 1 h of heat treatment at 500 °C. Further treatment led to a slower decrease in differential absorbance, down to ~0.1. Thus, heat treatment at 500 °C induced out-diffusion of silver ions to the glass surface with formation of SNIF, resulting in a 95% reduction in the total number of silver ions in the PTR-M glass sample compared to the nontreated ion-exchanged sample.

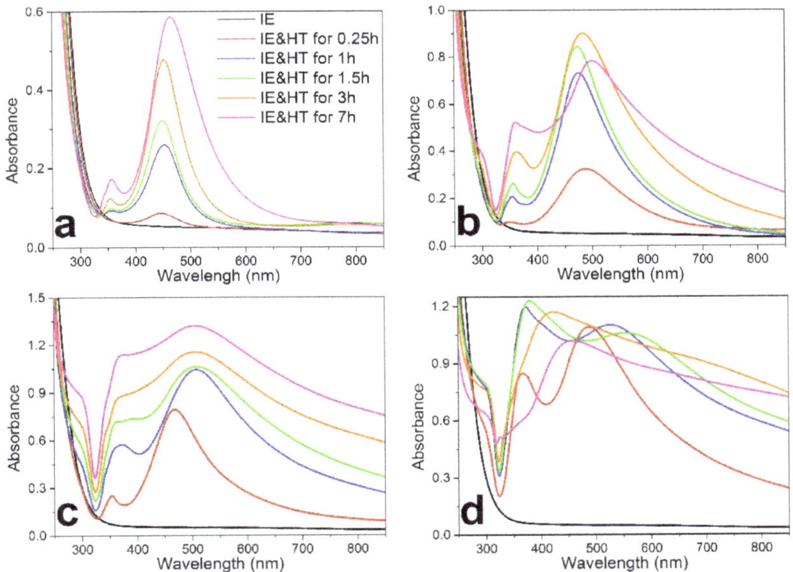

Figure 3. Absorbance spectra of PTR-M glass sample after the Na^+–Ag^+ IE and with SNIF obtained at 350 °C (**a**), 400 °C (**b**), 450 °C (**c**), and 500 °C (**d**) for different durations, indicated by colors.

XRD analysis of the PTR glass samples heat-treated at different temperatures for 3 h revealed characteristic peaks for metallic silver (Figure 5), confirming that the observed SNIF consisted of silver nanoparticles. SEM analysis of the of the ion-exchanged and heat-treated PTR-M glass surface revealed the presence of isolated nanometer and micrometer particles (Figure 6). SNIF formed at 350 °C consisted of nanoparticles of different shapes (triangle, square, and rectangular), while, at higher temperatures, nanoparticles had only spherical shape. The latter should be related to the melting of small SNPs during the heat treatment at temperatures above 400 °C [38–40]. Moreover, SEM analysis showed that the mean diameter of silver particles in the SNIF increased from 140 to 320 nm with heat treatment temperature increasing from 350 to 500 °C. This result correlates with the changes in absorbance spectra presented in Figure 3. The phase retardation of the electromagnetic field inside large particles led to the plasmon modes of higher multipole orders (quadrupolar, octupolar, and hexadecapolar) observed in the spectra [41].

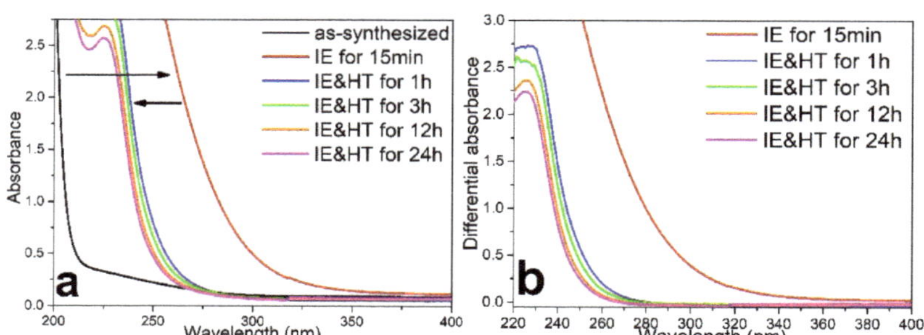

Figure 4. (**a**) Absorbance spectra of as-synthesized and Na^+–Ag^+ ion-exchanged PTR-M glass samples after heat treatment at 500 °C for different durations measured after removing the SNIF. (**b**) The differential absorbance spectra obtained by subtracting the spectrum of the as-synthesized sample from the others.

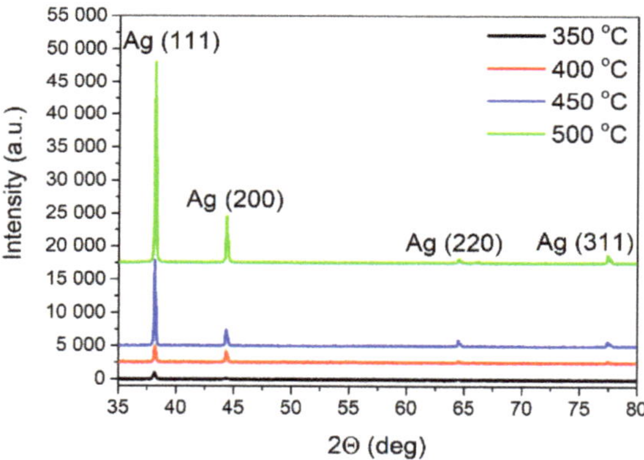

Figure 5. XRD spectra of the ion-exchanged and heat-treated for 3 h PTR-M glass samples with SNIF.

Considering possible mechanisms of the spontaneous growth of SNIF on the ion-exchanged PTR glass surface, the main question concerns the source of electrons for reducing silver ions. The mechanism of silver ion reduction by water vapor in a wet argon atmosphere responsible for growth of SNIF on the surface of Na^+–Ag^+ ion-exchanged borosilicate glass was suggested by Kaganovskii et al. [42]. The authors proposed silver–proton exchange with the formation of silver oxide that thermally decomposes further to silver and oxygen.

$$\equiv Si - O - Ag^+ + \frac{1}{2}H_2O \leftrightarrow \equiv Si - O - H^+ + \frac{1}{2}Ag_2O. \tag{1}$$

$$Ag_2O \rightarrow 2Ag^0 + O. \tag{2}$$

Other routes of silver reduction by glass impurities or nonbridging oxygen can be excluded considering that they would lead to coloration due to growth of SNPs inside the glass [43]. Thus, we suggest that SNIF on the surface of Na^+–Ag^+ ion-exchanged PTR glass surface growth is the result of interaction with the water molecules always present in air.

However, further research is needed to clarify the exact mechanism responsible for the growth of SNIF.

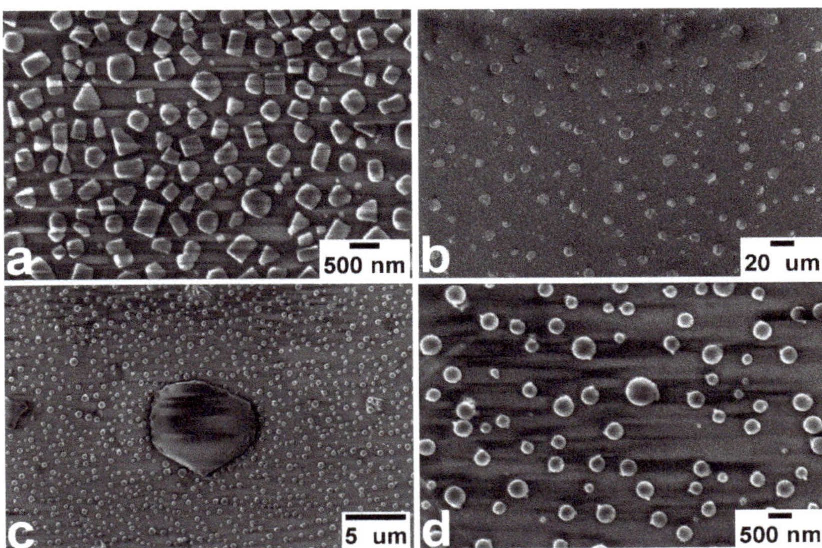

Figure 6. SEM image of the ion-exchanged PTR-M glass sample heat-treated at 350 °C (**a**) and 450 °C (**b–d**; various magnifications) for 3 h.

A drop (~5 mm in diameter) of 8×10^{-7} mol/L R6G aqueous solution was deposited on the samples to investigate SERS-active properties of silver films on the surface of the PTR-M glass. The Raman spectra were measured after drying the R6G drop with the confocal Raman microscope (Renishaw); a red laser beam (633 nm) was focused at the sample surface with a $50\times/0.55$ micro-objective. A significantly enhanced spectrum was observed at surface with SNIF compared to pristine PTR-M glass (Figure 7a). As growth of SNIF on the PTR glass is a diffusion process, it can be grown using the same substrate in several cycles of thermal treatment and mechanical removal. This makes Na^+–Ag^+ ion-exchanged PTR glass functionalized with SNIF an attractive platform for developing reusable SERS sensors. Absorbance spectra of the Na^+–Ag^+ ion-exchanged PTR-M glass with SNIF grown during several cycles of the heat treatment at 500 °C for 15 min and removing the film are shown on Figure 7b.

Thus, heat treatment at temperatures 350–500 °C of Na^+–Ag^+ ion-exchanged PTR-M glass with no dopants resulted in growth of SERS-active SNIF on the glass surface. Spectroscopic properties of SNPs on the glass surface can be controlled by changing the heat treatment temperature and time. Na^+–Ag^+ ion-exchanged PTR glass can be used as a substrate for developing reusable biochemical sensors based on the SERS phenomenon.

3.2. Photo- and Thermochromic Behavior of Na^+–Ag^+ Ion-Exchanged Br-Doped Glasses

In contrast to the PTR-M matrix glass with no dopants, Na^+–Ag^+ ion-exchanged bromine-containing glass PTR-Br (see Table 2) revealed an additional long-wavelength shift of the UV edge of strong absorption after the heat treatment at 500 °C (Figure 8a). Silver bromide crystals are known to possess strong absorption in the UV and visible range, with absorption coefficients as high as 1000 cm^{-1} at 400 nm, which may, thus, cause the long-wavelength shift [44]. Formation of AgBr nanocrystals in the ion-exchanged and heat-treated PTR-Br glass samples was confirmed by XRD analysis. XRD peaks related to silver bromide were clearly observed for all ion-exchanged and heat-treated samples (Figure 8b).

The mean diameter of AgBr nanocrystals calculated according to the well-known Scherrer equation increased from 5 to 10 nm upon extending the IE process from 15 min to 21 h.

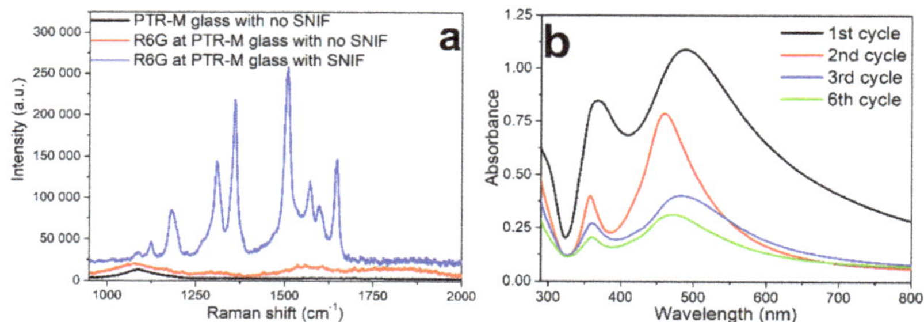

Figure 7. Raman spectra of PTR-M glass sample and R6G measured at ion-exchanged and heat-treated at 350 °C for 3 h (**a**). Absorbance spectra of PTR-M glass with SNIF after several cycles of the heat treatment at 500 °C for 15 min and removing the film before the following heat treatment at the same parameters (**b**).

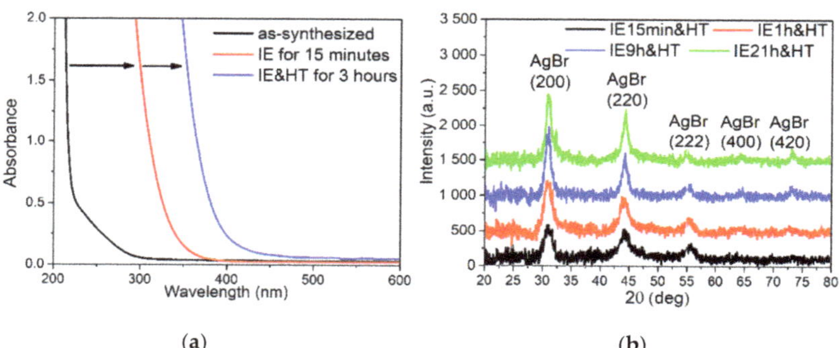

(**a**)   (**b**)

Figure 8. (**a**) Absorbance spectra of PTR-Br glass samples before and after the IE and heat treatment at 500 °C for 3 h. (**b**) XRD patterns of PTR-Br glass samples ion-exchanged for different time and heat-treated at 500 °C for 9 h.

Silver bromide is known to decompose to silver and bromine under UV or visible light.

$$AgBr \overset{h\nu}{\Rightarrow} Ag^0 + Br^0. \tag{3}$$

The process is used in photographic films; after exposure and decomposition of AgBr, silver atoms aggregate further into colloidal particles and form a latent image, while the produced bromine volatilizes. In the case of a glass matrix, bromine atoms are unable to leave the matrix, which enables multiple cycles of recording and relaxation of color centers. This principle underlies photochromic glasses that reversibly changes transmission depending on the illumination [45]. In 1968, Garfinkel showed that photochromic glasses can be produced by Na^+–Ag^+ ion exchange of borosilicate glasses doped with copper and bromine/chlorine [46]. Elliptical silver nanoparticles are known to be color centers in conventional photochromic glasses [47]. The experimental absorbance spectra consist of two broad bands corresponding to longitudinal and transverse plasmonic modes of elliptical SNPs. In order to study the photosensitivity of ion-exchanged PTR-Br glass, some samples were exposed to a mercury UV lamp (λ_{max} = 365 nm). However, only the sample ion-exchanged for 21 h showed a noticeable change in absorbance (Figure 9).

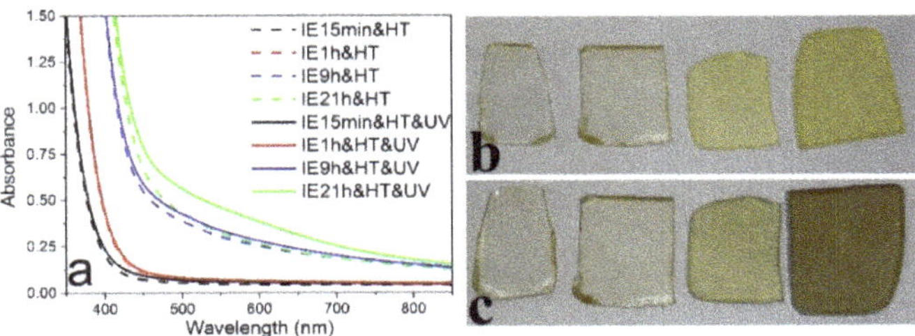

Figure 9. (**a**) Absorbance spectra of the ion-exchanged and heat-treated at 500 °C for 9 h PTR-Br glass samples before and after UV irradiation with a mercury lamp. (**b**) Photo of the samples before and (**c**) after the UV irradiation.

To enhance the formation of color centers in PTR-Br glass samples with AgBr nanocrystals, the samples were irradiated with a pulsed third-harmonic YAG:Nd laser (λ = 355 nm, f = 10Hz, pulse energy 20 mJ, pulse duration 7 ns). Intense laser UV light resulted in the growth of SNPs and, as a consequence, origin of a broad absorbance band in the visible range (Figure 10). Coloration and bleaching kinetics for the studied PTR-Br samples are shown in Figure 11. The amplitude of UV-induced absorbance increased with the Na^+–Ag^+ IE time as the concentration of the embedded silver ions determined concentration and distribution of AgBr nanocrystals formed during the subsequent heat treatment. It should be noted that saturation of coloration was observed already after 100 s of UV irradiation for samples ion-exchanged for 0.25, 1, and 9 h, while the sample ion-exchanged for 21 h did not show saturation even after 1000 s of UV irradiation.

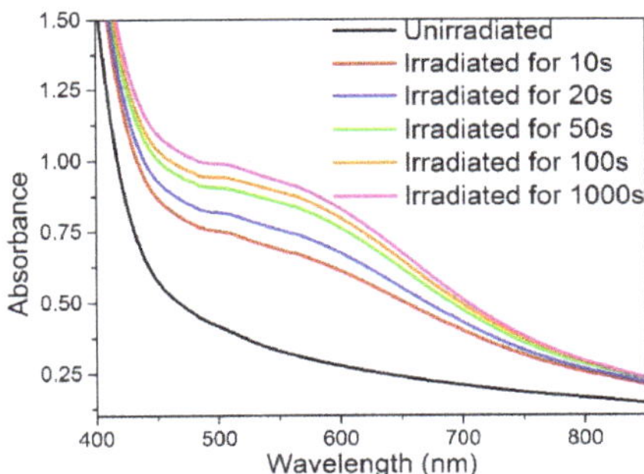

Figure 10. Absorbance spectra of the PTR-Br glass sample ion-exchanged for 9 h and heat-treated at 500 °C for 9 h after irradiation with UV pulses for different times.

After coloration by the UV pulsed laser, samples were kept at room temperature to study bleaching kinetics. Only partial bleaching of the samples was observed, while a number of color centers were stable at room temperature. For the samples ion-exchanged for 0.25, 1, and 9 h, the absorbance increment reached 20–30% of the peak value after 72 h of relaxation. At the same time, the sample ion-exchanged for 21 h possessed a 45% decrease in absorbance.

The observed differences in coloration and bleaching kinetics of the ion-exchanged samples with AgBr nanocrystals may be explained by the different thickness of the glass layer doped with the nanocrystals and by size effect. According to Araujo, the mean size of silver halide nanocrystals in photochromic glasses is 8–18 nm because smaller particles exhibit low coloration efficiency [47]. Thus, the low saturation level in the samples ion-exchanged for 15 min and 1 h can be explained by small AgBr nanocrystals with a mean size of 5 and 6 nm, respectively, calculated according to the Scherrer equation. Partial bleaching of the studied samples resulted from the absence of polyvalent ions, for example, copper, that participate in redox reactions occurring in conventional photochromic glasses during coloration and bleaching [45,47]. It should be noted that full relaxation of color centers induced by a UV mercury lamp and UV pulsed laser was observed after heating PTR-Br glass samples above 150 °C for 10–15 min.

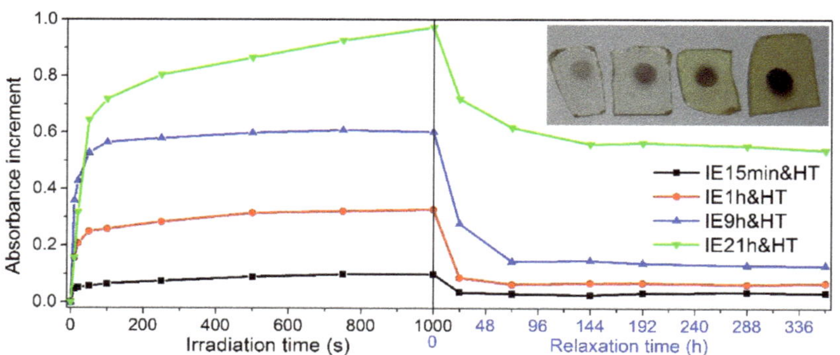

Figure 11. Coloring and bleaching kinetics for PTR-Br glass sample ion exchanged for different time and heat-treated at 500 °C for 9 h. Inset: photo of the samples after irradiation by the UV pulsed laser during 1000 s.

Silver bromide is an indirect semiconductor with direct and indirect band gap energies $E_{g,dir}$ = 4.276 eV and $E_{g,ind}$ = 2.713 eV for bulk crystals at cryogenic temperatures [48]. The energy bandgap of a semiconductor depends on temperature and decreases with its increase. In situ measurements of absorbance spectra revealed that the UV absorption edge of silver bromide nanocrystals formed in PTR-Br glass shifted to a greater wavelength with heating the sample (Figure 12). The effect was manifested as colored PTR-Br glass samples due to increasing absorption in the blue range (Figure 12, inset). After cooling the sample to room temperature and removing SNIF formed during the heating, the spectrum returned to the initial state. Figure 13a shows the dependence of the location of the UV absorption edge on the temperature of the samples ion-exchanged for different time. The experimental data can be fitted using a straight line for all samples. Moreover, for a wide temperature range between 300 and 500 °C, absorbance at a fixed wavelength linearly depends on the temperature for samples ion-exchanged for 1 and 9 h (Figure 13b). Thus, the thermochromic properties of PTR-Br glass samples containing AgBr nanocrystals can be used as temperature sensors or an eye-readable overheating indicator. It should be noted that chlorine-doped PTR glass samples with AgCl nanocrystals formed with the Na^+–Ag^+ IE were also studied. However, the photo- and thermochromic properties of PTR glass samples with AgCl nanocrystals was poorer compared to their bromine-doped counterparts.

3.3. Luminescent Properties of Silver Molecular Clusters Formed in Ce- and Sb-Doped Glasses

Silver molecular clusters (SMCs) are subnanometer silver aggregates that consist of several atoms/ions. SMCs are characterized by photoluminescence and do not possess plasmon resonance. Stabilized in different matrices, including liquids, polymers, glasses, and zeolites, SMCs are used for developing white LEDs [49], sensors [50], spectral converters [51,52], and data storage devices [53]. Glasses doped with SMCs attract particular attention due to transparency in a wide spectral range, high thermal stability, and excellent chemical durability. A PTR glass matrix transparent in the UV range opens the possibility to study the spectral properties of SMCs in details. To promote the growth of SMCs in the Na^+–Ag^+ ion-exchanged PTR glass, silver ions should be reduced to an atomic state. Cerium and antimony are polyvalent ions that can be in different charge states (Ce^{3+}/Ce^{4+} and Sb^{3+}/Sb^{5+}), which determines their spectroscopic manifestation in glass matrices [54,55]. These ions are typically used in PTR glasses to provide photosensitivity of the glass [3,4]. Possible redox reactions involving silver and cerium/antimony ions are as follows:

$$Ag^+ + Ce^{3+} \leftrightarrow Ag^0 + Ce^{4+}, \tag{4}$$

$$2Ag^+ + Sb^{3+} \leftrightarrow 2Ag^0 + Sb^{5+}. \tag{5}$$

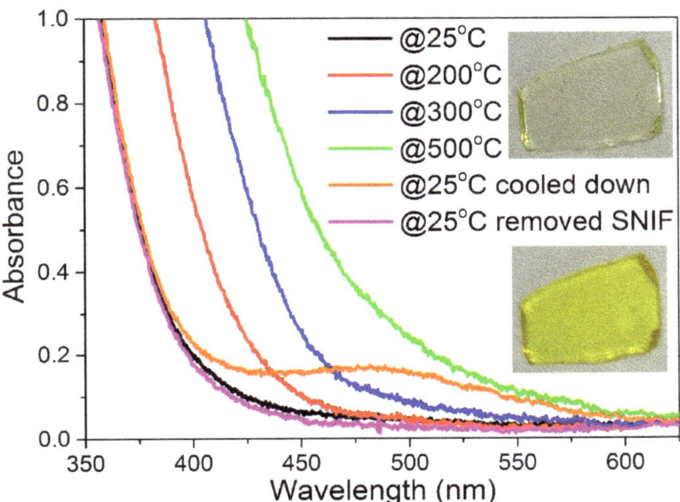

Figure 12. In situ absorbance spectra measured at different temperatures of PTR-Br glass sample ion-exchanged for 15 min and heat-treated at 500 °C for 9 h. Inset: photos of the sample at room temperature (**top image**) and after pulling it out from a muffle furnace heated to 500 °C (**bottom image**).

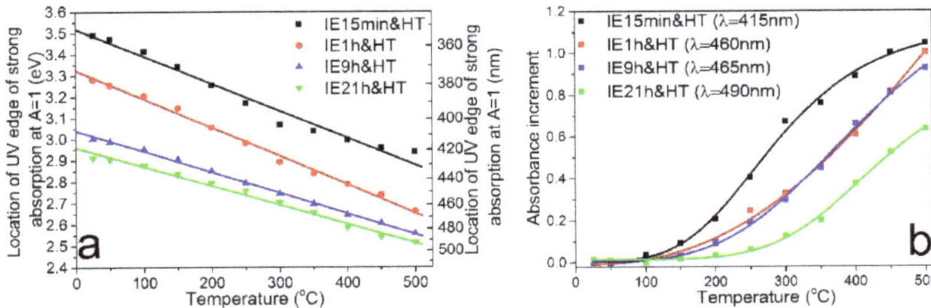

Figure 13. (**a**) Location of UV absorption edge and (**b**) absorbance increment at a fixed wavelength vs. temperature for PTR-Br glass samples ion-exchanged for different time and heat-treated at 500 °C for 9 h.

To determine the direction of a redox reaction, one needs to know the value of the standard electrode potential. For example, the reaction in Equation (4) proceeds in standard conditions from right to left. Adding silver into the PTR glass matrix was earlier shown to result in a decrease in the Ce^{4+}-related absorption band, proving that silver plays the role of a reducing agent for cerium ions in the PTR glass melt [56]. However, the concentration of Ag^+ ions in the ion-exchanged PTR glass layer exceeds by orders the equilibrium value that can be reached via glass synthesis, whereby the equilibria in Equations (4) and (5) shift to the right-hand side despite the direction of the reaction in standard conditions. Neutral silver atoms obtained as a result of the redox reactions can aggregate further in SMCs or

NPs depending on their size. As shown below, the spectral luminescent properties of PTR Na$^+$-Ag$^+$ ion-exchanged glass can be tuned in a wide range by varying the concentration of glass dopants, as well as the IE and subsequent heat treatment parameters.

The concentration of polyvalent ions determines the rate of the redox reaction with silver. Thus, a higher content of polyvalent ions allows a higher concentration of SMCs and SNPs to be obtained in glass at fixed parameters of the IE and heat treatment processes (Figure 14). It is worth noting that, for developing efficient phosphors or spectral converters based on a glass doped with SMCs, the concentration of polyvalent ions should be very low, so as to increase the possibility of SMCS growth consisting only of few atoms. In the other case, the growth of large clusters and/or SNPs results in luminescence quenching and a significant lowering of the photoluminescence quantum yield (PLQY). For example, a PLQY value as high as 65%, which is among the highest reported values for glasses doped with SMCs [36], was achieved in PTR-Sb2 glass containing only 0.002 mol.% of Sb$_2$O$_3$. An increase in Sb$_2$O$_3$ content to 0.01 mol.% led to almost a twofold decrease in photoluminescence quantum yield.

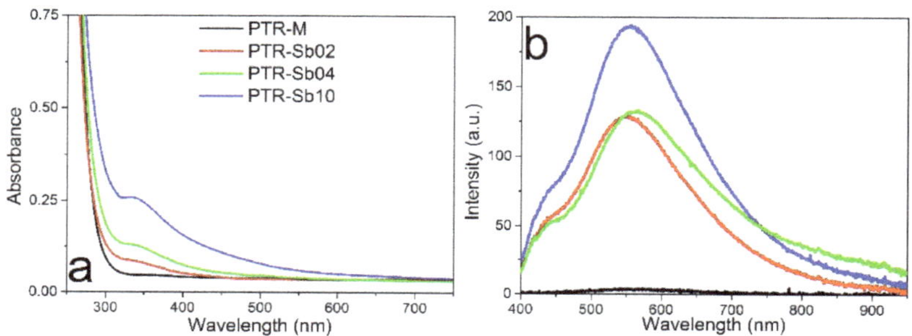

Figure 14. Absorbance (**a**) and emission (**b**) spectra of undoped PTR-M and antimony-doped PTR glass samples ion-exchanged for 15 min and heat-treated at 450 °C for 15 h. The excitation wavelength was 350 nm.

Heat treatment at temperatures below the glass transition temperature was found to result in bright photoluminescence due to the growth of SMCs. PLQY magnitude increases with heat treatment temperature up to 450 °C. Heat treatment at 500 °C, i.e., above the T$_g$ value of the studied antimony-doped glass samples, led to the appearance of an absorption band around 420 nm corresponding to the localized surface plasmon resonance of SNPs (Figure 15a). The growth of SNPs inevitably caused luminescence quenching due to the absorption of SMC emission and the decrease in concentration of emission centers due to the transformation of SMCs into SNPs (Figure 15b).

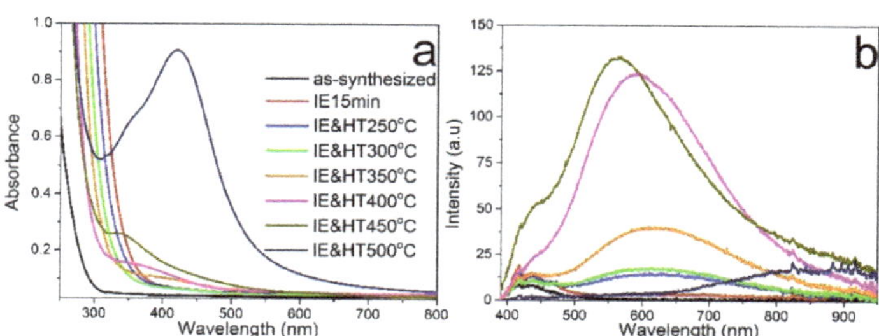

Figure 15. Absorbance (**a**) and emission spectra (**b**) of PTR-Sb10 glass samples as-synthesized, ion-exchanged, and heat-treated after the IE process at different time for 15 h. The excitation wavelength was 350 nm.

IE parameters, including silver content in the salt melt, temperature, and time, can also be used to tune the optical properties of silver nanostructures in Na^+–Ag^+ ion-exchanged glasses. The IE time and temperature influence the diffusion rate and, as a result, the thickness of the silver-enriched glass layer. Thus, these parameters change the properties of the obtained silver nanostructures in a similar way. It was found that varying IE time allows tuning the location of SMC emission in a wide spectral range. This opens up prospects for developing highly efficient, thermostable, solid-state LEDs based on PTR glasses with SMCs. The effect of the IE time on the absorbance and emission of SMCs is demonstrated in Figures 16 and 17. Emission color varied from cool white to warm and orange upon increasing the IE time (Figure 16c). It should be noted that the PLQY of the samples decreased from 60% to 15% with the extension of IE duration [37]. This effect is related to the increase in absorbance of the remaining silver ions at the excitation wavelength with an increase in the IE time (a longer IE process leads to a higher amount of silver remaining in ionic form in the PTR glass due to the constant concentration of reducing agent). However, we believe that PLQY values can be substantially increased by using a salt melt with low silver content. The silver concentration in subsurface layers of the Na^+-Ag^+ ion-exchanged glass is known to nonlinearly depend on the silver content in the salt melt for most glasses. Typically, the surface concentration of silver grows exponentially with $AgNO_3$ content below 5 mol.% and saturates at greater values [57]. IE with a low silver content may allow decreasing the absorption of silver ions that remain in the PTR glass after the heat treatment.

The presented results show that the photoluminescence properties of SMCs in the PTR glass matrix can be tuned in a wide range by controlling the concentration of reducing agents, as well as the parameters of the Na^+–Ag^+ IE and subsequent heat treatment processes. The wide emission spectrum of SMCs in the ion-exchanged PTR glass matrix and the high PLQY magnitude pave the way for applications of such glasses in lighting as phosphors for white LEDs and in photovoltaics as spectral converters.

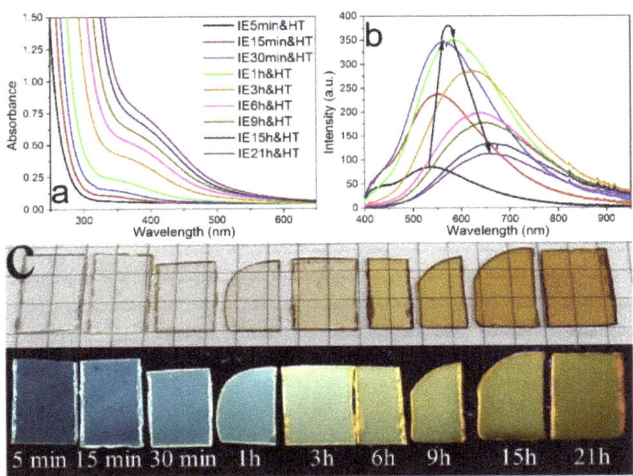

Figure 16. Absorbance (**a**) and emission (**b**) spectra of PTR-Sb2 glass samples ion-exchanged for different time and heat-treated at 450 °C for 15 h. The excitation wavelength was 350 nm. (**c**) Photo of the same samples in daylight (**top row**) and UV light (**bottom row**).

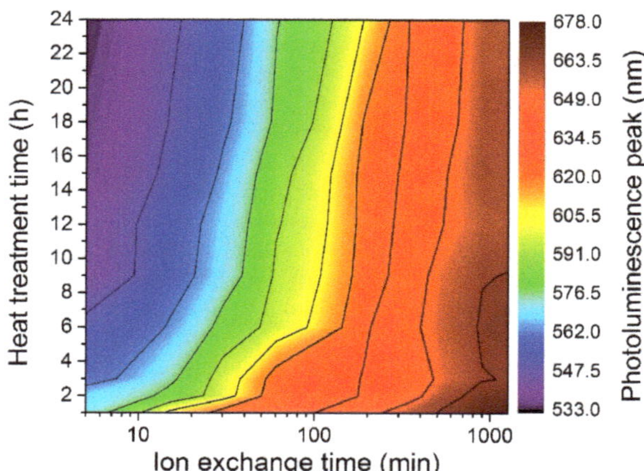

Figure 17. Dependence of photoluminescence peak location of PTR-Sb2 glass samples on the time of Na$^+$–Ag$^+$ IE and subsequent heat treatment at 450 °C.

4. Conclusions and Perspectives

In this work, the great effect of small additives in sodium–zinc–alumina glass composition on the spectral and optical properties of silver nanostructures grown on the surface or subsurface layers of Na$^+$–Ag$^+$ ion-exchanged glasses was demonstrated. Silver ions introduced in the PTR glass matrix containing no polyvalent ions can interact with the atmosphere water molecules during heat treatment to form SERS-active silver nanoisland films on the glass surface. Trivalent cerium and antimony ions typically used as photosensitive additives in PTR glasses were shown to act as reducing agents for silver ions in Na$^+$–Ag$^+$ ion-exchanged glasses and to promote the formation of luminescent silver molecular clusters. The concentration of polyvalent ions, as well as the ion exchange and subsequent heat treatment parameters, can be used to tune and optimize the emission of silver molecular clusters in a wide range.

Na$^+$–Ag$^+$ ion exchange of Br(Cl)-containing PTR glass followed by heat treatment above the glass transition temperature led to the growth of silver bromide nanocrystals. The obtained nano-glass ceramic possesses photochromic and thermochromic properties. The presented experimental findings indicate that small additives contained in glass play a key role in the chemical reactions with silver ions introduced by Na$^+$–Ag$^+$ ion exchange and determine the properties of the obtained silver nanostructures and functionality of nano-glass ceramics. We believe that the results of this work can be useful for developing biochemical sensors based on the SERS phenomenon, white LEDs, spectral converters, and temperature sensors.

Author Contributions: Conceptualization, Y.S. and N.N.; formal analysis, Y.S.; funding acquisition, N.N.; investigation, Y.S.; methodology, Y.S. and A.I.; project administration, N.N.; validation, Y.S., N.N. and A.I.; writing—original draft, Y.S.; writing—review and editing, N.N. and A.I. All authors have read and agreed to the published version of the manuscript.

Funding: This work was financially supported by the Russian Science Foundation (Project # 20-19-00559).

Institutional Review Board Statement: Not applicable.

Informed Consent Statement: Not applicable.

Data Availability Statement: Data is contained within the article.

Acknowledgments: The authors are thankful to Oreshkina K.S. for XRD measurements and Klyukin D.A. for SEM images.

Conflicts of Interest: The authors declare no conflict of interest.

References

1. Nikonorov, N.; Aseev, V.; Dubrovin, V.; Ignatiev, A.; Ivanov, S.; Sgibnev, Y.; Sidorov, A. Photonic, Plasmonic, Fluidic, and Luminescent Devices Based on New Polyfunctional Photo-Thermo-Refractive Glass. In *Optics, Photonics and Laser Technology*; Ribeiro, P., Raposo, M., Eds.; Springer International Publishing: Berlin/Heidelberg, Germany, 2018; p. 246, ISBN 978-3-319-98547-3.
2. Nikonorov, N.; Ivanov, S.; Dubrovin, V.; Ignatiev, A. New Photo-Thermo-Refractive Glasses for Holographic Optical Elements: Properties and Applications. In *Holographic Materials and Optical Systems*; Naydenova, I., Ed.; IntechOpen: London, UK, 2017; p. 503.
3. Lumeau, J.; Zanotto, E.D. A review of the photo-thermal mechanism and crystallization of photo-thermo-refractive (PTR) glass. *Int. Mater. Rev.* **2017**, *62*, 348–366. [CrossRef]
4. Ivanov, S.; Dubrovin, V.; Nikonorov, N.; Stolyarchuk, M.; Ignatiev, A. Origin of refractive index change in photo-thermo-refractive glass. *J. Non. Cryst. Solids* **2019**, *521*, 119496. [CrossRef]
5. Sato, Y.; Taira, T.; Smirnov, V.; Glebova, L.; Glebov, L. Continuous-wave diode-pumped laser action of Nd^{3+}-doped photo-thermo-refractive glass. *Opt. Lett.* **2011**, *36*, 2257–2259. [CrossRef]
6. Ivanov, S.A.; Lebedev, V.F.; Ignat'ev, A.I.; Nikonorov, N.V. Laser action on neodymium heavily doped photo-thermo-refractive glass. In Proceedings of the 1st International Symposium on Advanced Photonic Materials, St. Petersburg, Russia, 27 June–1 July; ITMO University: St. Petersburg, Russia, 2016; pp. 29–31.
7. Sgibnev, Y.; Nikonorov, N.; Ignatiev, A.; Vasilyev, V.; Sorokina, M. Photostructurable photo-thermo-refractive glass. *Opt. Express* **2016**, *24*, 4563. [CrossRef]
8. Sgibnev, Y.M.; Nikonorov, N.V.; Vasilev, V.N.; Ignatiev, A.I. Optical Gradient Waveguides in Photo-Thermo-Refractive Glass Formed by Ion Exchange Method. *J. Light. Technol.* **2015**, *33*, 3730–3735. [CrossRef]
9. Stepanov, A.L.; Hole, D.E.; Townsend, P.D. Formation of silver nanoparticles in soda-lime silicate glass by ion implantation near room temperature. *J. Non. Cryst. Solids* **1999**, *260*, 65–74. [CrossRef]
10. Arnold, G.W.; Borders, J.A. Aggregation and migration of ion-implanted silver in lithia-alumina-silica glass. *J. Appl. Phys.* **1977**, *48*, 1488–1496. [CrossRef]
11. Onbaşlı, M.C.; Tandia, A.; Mauro, J.C. Mechanical and compositional design of high-strength Corning Gorilla® Glass. In *Handbook of Materials Modeling: Applications: Current and Emerging Materials*; Springer International Publishing: New York, NY, USA, 2020; pp. 1997–2019.
12. Findakly, T. Glass waveguides by ion exchange: A review. *Opt. Eng.* **1985**, *25*, 244–250. [CrossRef]
13. Tervonen, A.; West, B.R.; Honkanen, S. Ion-exchanged glass waveguide technology: A review. *Opt. Eng.* **2011**, *50*, 71107. [CrossRef]
14. Karlsson, S.; Jonson, B.; Stålhandske, C. The technology of chemical glass strengthening—A review. *Glas. Technol. Eur. J. Glas. Sci. Technol. Part A* **2010**, *51*, 41–54.
15. Haes, A.J.; Van Duyne, R.P. A nanoscale optical biosensor: Sensitivity and selectivity of an approach based on the localized surface plasmon resonance spectroscopy of triangular silver nanoparticles. *J. Am. Chem. Soc.* **2002**, *124*, 10596–10604. [CrossRef] [PubMed]
16. Fan, M.; Brolo, A.G. Silver Nanoparticles Self Assembly as SERS Substrates with Near Single Molecule Detection Limit. *Phys. Chem. Chem. Phys.* **2009**, *11*, 7381–7389. [CrossRef]
17. Enrichi, F.; Quandt, A.; Righini, G.C. Plasmonic enhanced solar cells: Summary of possible strategies and recent results. *Renew. Sustain. Energy Rev.* **2018**, *82*, 2433–2439. [CrossRef]
18. Ivanova, N.; Gugleva, V.; Dobreva, M.; Pehlivanov, I.; Stefanov, S.; Andonova, V. Silver Nanoparticles as Multi-Functional Drug Delivery Systems. In *Nanomedicines*; IntechOpen: London, UK, 2018.
19. Guo, H.; Feng, M.; Song, F.; Li, H.; Ren, A.; Wei, X.; Li, Y.; Xu, X. Q-switched Erbium-doped Fiber Laser Based on Silver Nanoparticles as a Saturable Absorber. *IEEE Photonics Technol. Lett.* **2015**, *28*, 135–138. [CrossRef]
20. Sarina, S.; Waclawik, E.R.; Zhu, H. Photocatalysis on Supported Gold and Silver Nanoparticles under Ultraviolet and Visible Light Irradiation. *Green Chem.* **2013**, *17*, 1814–1833. [CrossRef]
21. Fleischmann, M.; Hendra, P.J.; McQuillan, A.J. Raman spectra of pyridine adsorbed at a silver electrode. *Chem. Phys. Lett.* **1974**, *26*, 163–166. [CrossRef]
22. Doering, W.E.; Nie, S. Single-Molecule and Single-Nanoparticle SERS: Examining the Roles of Surface Active Sites and Chemical Enhancement. *J. Phys. Chem. B* **2002**, *106*, 311–317. [CrossRef]
23. Kneipp, J.; Kneipp, K. SERS—A single-molecule and nanoscale tool for bioanalytics. *Chem. Soc. Rev.* **2008**, *37*, 1052–1060. [CrossRef]
24. Chen, Y.; Karvonen, L.; Säynätjoki, A.; Ye, C.; Tervonen, A.; Honkanen, S. Ag nanoparticles embedded in glass by two-step ion exchange and their SERS application. *Opt. Mater. Express* **2011**, *1*, 164. [CrossRef]
25. Chen, Y.; Jaakola, J.J.; Säynätjoki, A.; Tervonen, A.; Honkanen, S. Glass-embedded silver nanoparticle patterns by masked ion-exchange process for surface-enhanced Raman scattering. *J. Raman Spectrosc.* **2011**, *42*, 936–940. [CrossRef]
26. Zhurikhina, V.V.; Brunkov, P.N.; Melehin, V.G.; Kaplas, T.; Svirko, Y.; Rutckaia, V.V.; Lipovskii, A.A. Self-assembled silver nanoislands formed on glass surface via out-diffusion for multiple usages in SERS applications. *Nanoscale Res. Lett.* **2012**, *7*, 676. [CrossRef]
27. Chervinskii, S.; Sevriuk, V.; Reduto, I.; Lipovskii, A. Formation and 2D-patterning of silver nanoisland film using thermal poling and out-diffusion from glass. *J. Appl. Phys.* **2013**, *114*, 224301. [CrossRef]

28. Heisler, F.; Babich, E.; Scherbak, S.; Chervinskii, S.; Hasan, M.; Samusev, A.; Lipovskii, A.A. Resonant Optical Properties of Single Out-Diffused Silver Nanoislands. *J. Phys. Chem. C* **2015**, *119*, 26692–26697. [CrossRef]
29. Babich, E.S.; Redkov, A.V.; Reduto, I.V.; Scherbak, S.A.; Kamenskii, A.N.; Lipovskii, A.A. Raman enhancement by individual silver hemispheroids. *Appl. Surf. Sci.* **2017**, *397*, 119–124. [CrossRef]
30. Cattaruzza, E.; Mardegan, M.; Trave, E.; Battaglin, G.; Calvelli, P.; Enrichi, F.; Gonella, F. Modifications in silver-doped silicate glasses induced by ns laser beams. *Appl. Surf. Sci.* **2011**, *257*, 5434–5438. [CrossRef]
31. Ajami, A.; Husinsky, W.; Svecova, B.; Vytykacova, S.; Nekvindova, P. Saturable absorption of silver nanoparticles in glass for femtosecond laser pulses at 400 nm. *J. Non. Cryst. Solids* **2015**, *426*, 159–163. [CrossRef]
32. Sgibnev, E.M.; Ignatiev, A.I.; Nikonorov, N.V.; Efimov, A.M.; Postnikov, E.S. Effects of silver ion exchange and subsequent treatments on the UV-VIS spectra of silicate glasses. I. Undoped, CeO2-doped, and (CeO 2 + Sb2O3)-codoped photo-thermo-refractive matrix glasses. *J. Non. Cryst. Solids* **2013**, *378*, 213–226. [CrossRef]
33. Kinnan, M.K.; Chumanov, G. Plasmon Coupling in Two-Dimensional Arrays of Silver Nanoparticles: II. Effect of the Particle Size and Interparticle Distance. *J. Phys. Chem. C* **2010**, *114*, 7496–7501. [CrossRef]
34. Spierings, G. Optical absorption of Ag+ ions in 11(Na, Ag)2O·11B2O3·78SiO2 glass. *J. Non. Cryst. Solids* **1987**, *94*, 407–411. [CrossRef]
35. Efimov, A.M.; Ignatiev, A.I.; Nikonorov, N.V.; Postnikov, E.S. Ultraviolet-VIS spectroscopic manifestations of silver in photo-thermo-refractive glass matrices. *Eur. J. Glas. Sci. Technol. Part A* **2013**, *54*, 155–164.
36. Sgibnev, Y.M.; Nikonorov, N.V.; Ignatiev, A.I. High efficient luminescence of silver clusters in ion-exchanged antimony-doped photo-thermo-refractive glasses: Influence of antimony content and heat treatment parameters. *J. Lumin.* **2017**, *188*, 172–179. [CrossRef]
37. Sgibnev, Y.; Asamoah, B.; Nikonorov, N.; Honkanen, S. Tunable photoluminescence of silver molecular clusters formed in Na+-Ag+ ion-exchanged antimony-doped photo-thermo-refractive glass matrix. *J. Lumin.* **2020**, *226*, 117411. [CrossRef]
38. Luo, W.; Hu, W.; Xiao, S. Size effect on the thermodynamic properties of silver nanoparticles. *J. Phys. Chem. C* **2008**, *112*, 2359–2369. [CrossRef]
39. Asoro, M.A.; Damiano, J.; Ferreira, P.J. Size effects on the melting temperature of silver nanoparticles: In-situ TEM observations. *Microsc. Microanal.* **2009**, *15*, 706–707. [CrossRef]
40. Zhao, S.J.; Wang, S.Q.; Cheng, D.Y.; Ye, H.Q. Three distinctive melting mechanisms in isolated nanoparticles. *J. Phys. Chem. B* **2001**, *105*, 12857–12860. [CrossRef]
41. Kumbhar, A.S.; Kinnan, M.K.; Chumanov, G. Multipole plasmon resonances of submicron silver particles. *J. Am. Chem. Soc.* **2005**, *127*, 12444–12445. [CrossRef]
42. Kaganovskii, Y.; Mogilko, E.; Lipovskii, A.A.; Rosenbluh, M. Formation of nanoclusters in silver-doped glasses in wet atmosphere. *J. Phys. Conf. Ser.* **2007**, *61*, 508–512. [CrossRef]
43. Araujo, R. Colorless glasses containing ion-exchanged silver. *Appl. Opt.* **1992**, *31*, 5221. [CrossRef]
44. Tutihasi, S. Optical absorption by silver halides. *Phys. Rev.* **1954**, *105*, 882–884. [CrossRef]
45. Dotsenko, L.B.; Glebov, V.A.T. *Physics and Chemistry of Photochromic Glasses*; CRC Press: Boca Raton, FL, USA, 1998.
46. Garfinkel, H.M. Photochromic glass by silver ion exchange. *Appl. Opt.* **1968**, *7*, 789–794. [CrossRef]
47. Araujo, R.J. Photochromic Glass. *Treatise Mater. Sci. Technol.* **1977**, *12*, 91–122. [CrossRef]
48. Madelung, O. *Semiconductors: Data Handbook*; Springer: Berlin/Heidelberg, Germany, 2004; ISBN 9783662036488.
49. Kuznetsov, A.S.; Tikhomirov, V.K.; Moshchalkov, V.V. UV-driven efficient white light generation by Ag nanoclusters dispersed in glass host. *Mater. Lett.* **2013**, *92*, 4–6. [CrossRef]
50. Dubrovin, V.D.; Ignatiev, A.I.; Nikonorov, N.V.; Sidorov, A.I.; Shakhverdov, T.A.; Agafonova, D.S. Luminescence of silver molecular clusters in photo-thermo-refractive glasses. *Opt. Mater. Amst.* **2014**, *36*, 753–759. [CrossRef]
51. Cattaruzza, E.; Caselli, V.M.; Mardegan, M.; Gonella, F.; Bottaro, G.; Quaranta, A.; Valotto, G.; Enrichi, F. Ag+↔Na+ ion exchanged silicate glasses for solar cells covering: Down-shifting properties. *Ceram. Int.* **2015**, *41*, 7221–7226. [CrossRef]
52. Sgibnev, Y.; Cattaruzza, E.; Dubrovin, V.; Vasilyev, V. Photo-Thermo-Refractive Glasses Doped with Silver Molecular Clusters as Luminescence Downshifting Material for Photovoltaic Applications. *Part. Part. Syst. Charact.* **2018**, *35*, 1800141. [CrossRef]
53. Bourhis, K.; Royon, A.; Papon, G.; Bellec, M.; Petit, Y.; Canioni, L.; Dussauze, M.; Rodriguez, V.; Binet, L.; Caurant, D.; et al. Formation and thermo-assisted stabilization of luminescent silver clusters in photosensitive glasses. *Mater. Res. Bull.* **2013**, *48*, 1637–1644. [CrossRef]
54. Efimov, A.M.; Ignatiev, A.I.; Nikonorov, N.V.; Postnikov, E.S. Quantitative UV–VIS spectroscopic studies of photo-thermo-refractive glasses. II. Manifestations of Ce3+ and Ce(IV) valence states in the UV absorption spectrum of cerium-doped photo-thermo-refractive matrix glasses. *J. Non. Cryst. Solids* **2013**, *361*, 26–37. [CrossRef]
55. Ehrt, D. Photoluminescence in the UV-VIS region of polyvalent ions in glasses. *J. Non. Cryst. Solids* **2004**, *348*, 22–29. [CrossRef]
56. Efimov, A.M.; Ignatiev, A.I.; Nikonorov, N.V.; Postnikov, E.S. Photo-Thermo-Refractive Glasses: Effects of Dopants on Their Ultraviolet Absorption Spectra. *Int. J. Appl. Glas. Sci.* **2015**, *6*, 109–127. [CrossRef]
57. Zhurikhina, V.V.; Petrov, M.I.; Sokolov, K.S.; Shustova, O.V. Ion-exchange characteristics of sodium-calcium-silicate glass: Calculation from mode spectra. *Tech. Phys.* **2010**, *55*, 1447–1452. [CrossRef]

 *applied sciences*

Article

Investigation of Ytterbium Incorporation in Lithium Niobate for Active Waveguide Devices

Christian E. Rüter *, Dominik Brüske, Sergiy Suntsov and Detlef Kip

Faculty of Electrical Engineering, Helmut Schmidt University, 22043 Hamburg, Germany;
bruesked@hsu-hh.de (D.B.); suntsov@hsu-hh.de (S.S.); kip@hsu-hh.de (D.K.)
* Correspondence: ceh@hsu-hh.de

Received: 14 February 2020; Accepted: 11 March 2020; Published: 24 March 2020

Featured Application: Active waveguide devices, waveguide lasers.

Abstract: In this work, we report on an investigation of the ytterbium diffusion characteristics in lithium niobate. Ytterbium-doped substrates were prepared by in-diffusion of thin metallic layers coated onto x- and z-cut congruent substrates at different temperatures. The ytterbium profiles were investigated in detail by means of secondary neutral mass spectroscopy, optical microscopy, and optical spectroscopy. Diffusion from an infinite source was used to determine the solubility limit of ytterbium in lithium niobate as a function of temperature. The derived diffusion parameters are of importance for the development of active waveguide devices in ytterbium-doped lithium niobate.

Keywords: thermal diffusion; active waveguide devices; laser active materials

---

1. Introduction

The ferroelectric crystal lithium niobate ($LiNbO_3$) is a well-known material for various optical applications due to its favorable electro-optical, acousto-optical, piezoelectric, and nonlinear properties. Furthermore, low-loss waveguides can be implemented in rare-earth-doped $LiNbO_3$ via several fabrication techniques [1–6], leading to the development of waveguide amplifiers as well as waveguide lasers. In the past, a variety of efficient erbium- (Er) and neodymium (Nd)-doped $LiNbO_3$ waveguide lasers have been realized [7–10]. Another attractive laser material is $Yb:LiNbO_3$. Due to the simple energy level scheme of ytterbium (Yb) with only one excited state, a small laser quantum defect can be achieved, allowing for the development of highly efficient Yb lasers [11]. Another benefit of Yb is the large gain bandwidth, which allows for wide wavelength tuning and the generation of ultrashort pulses [12,13].

In 1995, Jones et al. reported the demonstration of an $Yb:LiNbO_3$ waveguide laser with a slope efficiency of 16% [14]. Due to photorefractive damage, the y-propagation Ti-diffused $Yb:LiNbO_3$ waveguide laser showed unstable laser operation. Later, Fujimura et al. reported continuous-wave lasing with a slope efficiency of 7% in z-cut $LiNbO_3$ annealed/proton-exchanged (APE) waveguides [15] by using thermally in-diffused Yb. While in the latter works, channel waveguides were used, another method which also reduces photorefractive damage is using ridge waveguides [16]. In such geometry, further improvement of the laser performance can be achieved due to the smaller mode fields and significantly improved overlap of optical pump and laser modes, when compared to their channel counterparts. Recently, a novel fabrication method has been reported comprising ridge definition by diamond blade dicing followed by three-side Er/Nd and Ti in-diffusion for the development of highly efficient rare-earth-doped $Ti:LiNbO_3$ ridge waveguide lasers [17–19]. This fabrication technique is a promising method for increasing the slope efficiency of $Yb:LiNbO_3$ waveguide lasers, too. For the development of low-loss $Yb:Ti:LiNbO_3$ ridge waveguides with optimized overlap of the Yb doping profile and the intensity profiles of the guided modes, it is essential to know the diffusion constants and the maximum solubility of Yb in LiNbO$_3$. Apart from a single value for the diffusivity at 1100 °C in z-direction [20], only a few data regarding the diffusion doping of LiNbO$_3$ with Yb can be found [14,21].

In this work, we report on an investigation of the incorporation of Yb into LiNbO$_3$ by in-diffusion. To determine the diffusion constants and the maximum solubility of Yb in LiNbO$_3$, diffusion profiles measured by SNMS (secondary neutral mass spectrometry) were analyzed. With these data, a first ridge waveguide sample was prepared, and the absorption and emission spectra of the in-diffused Yb ions were measured and compared with literature values.

2. Experimental Methods

2.1. Sample Preparation

For the sample fabrication, optical-grade x- and z-cut wafers of congruent LiNbO$_3$ (Yamaju Ceramics Co., LTD) were used. Pieces of $10 \times 10 \times 1$ mm^3 were prepared by diamond saw dicing. The samples were cleaned using successive baths of acetone, isopropanol and deionized water, and an ultrasonic cleaner. Plasma cleaning in oxygen was applied as the final cleaning step prior to vacuum deposition. The samples were coated with 20 nm thick layers of Yb, with a purity of 99.9%, using electron beam evaporation. For annealing, the samples were placed in a platinum crucible in a high temperature elevator furnace for diffusion time of 30 h at different temperatures between 930 and 1130 °C. The heating and cooling rates were constant for all samples.

One sample with dimensions of $1 \times 25 \times 8$ mm^3 (x y z) from the x-cut wafer was also used to prepare Yb-doped ridge waveguides. For this, ridges were prepared in the x surface running parallel to the y-axis by precise diamond blade milling, and these samples were coated with 22 nm of Yb under symmetric angles of ±60°. For diffusion, the coated ridges were annealed at $T = 1125$ °C for $t = 216$ h. Afterwards, 85 nm of Ti were deposited under symmetric angles of ±60° and in-diffused to prepare the ridge waveguides as described in [17].

2.2. SNMS Measurements

The SNMS measurements were performed using the Multimethod System with INA-X from SPECS. For the direct bombardment mode during sputtering, an RF-excited Krypton plasma was used. Samples were protected by a copper mask defining a free area with a diameter of 5 mm in the central part of each sample. After SNMS analysis, the sputtered crater was measured using a white light interferometer (WLI from FRT) to determine the sputtering rate. The rate was (0.45 ± 0.02) nm/s for all samples.

3. Results and Discussion

Figure 1 shows the depth profiles of ^{7}Li, ^{93}Nb, ^{16}O, and ^{70}Yb on a semilogarithmic scale giving two examples of SNMS spectra of the Yb-doped LiNbO$_3$ samples used in this analysis. Figure 1a demonstrates that at lower temperatures, a certain amount of Yb remains in form of a thin film on the surface, indicating that the Yb reservoir has not been exhausted.

Diffusion Theory

The Yb concentration profiles obtained in this work can be well approximated utilizing the one-dimensional diffusion theory described by Fick's second law of diffusion.

$$\frac{\partial C}{\partial t} = D\frac{\partial^2 C}{\partial y^2},$$ (1)

where D is the diffusion coefficient and y quantifies the direction normal to the surface. Assuming an inexhaustible source (i.e., a thick film) of Yb on the surface, the solution of Fick's law (1) can be approximated by

$$C(y) = C_{\max}\,\mathrm{erfc}(y/d_{erf}),\tag{2}$$

with $C_{\max}$ being the concentration of Yb at the crystal surface and the diffusion depth $d_{erf} = \sqrt{4Dt}$. In the thick-film diffusion regime. $C_{\max}$ describes the solubility of Yb in LiNbO$_3$. This value is given by

$$C_{\max} = \sqrt{\frac{\pi}{4Dt}}\int_0^\infty C(y)\,dy.\tag{3}$$

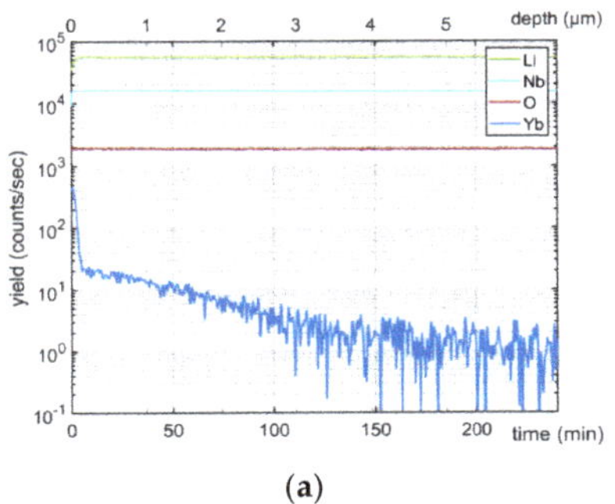

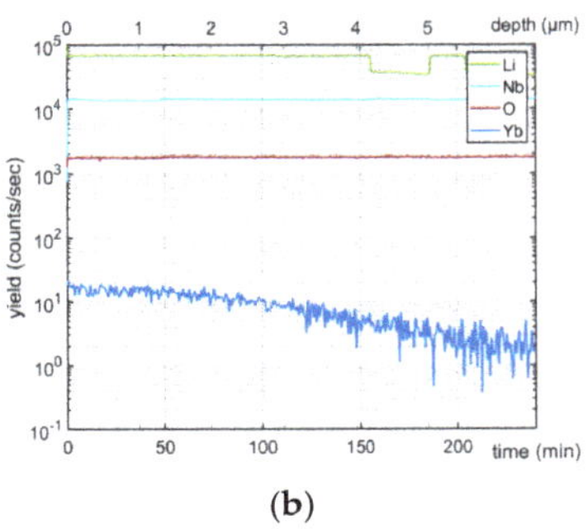

Figure 1. Secondary neutral mass spectrometry (SNMS) spectra of x-cut Yb:LiNbO$_3$. The two samples were coated with 20 nm Yb, and the diffusion parameters are (**a**) $T = 1060\,°\mathrm{C}$, $t = 30\,\mathrm{h}$ and (**b**) $T = 1130\,°\mathrm{C}$, $t = 30\,\mathrm{h}$, respectively.

Using the law of mass conservation, the time needed for the Yb source to be depleted can be calculated with the metallic Yb density ρ, the atomic mass m_{Yb}, and the thickness τ of the initial Yb layer

$$t_d = \left(\frac{\rho\tau}{2\,m_{\mathrm{Yb}}\,C_{\max}}\right)^2\frac{\pi}{D}.\tag{4}$$

For sufficiently thin films and/or sufficiently long in-diffusion times ($t > t_d$), the Yb source is depleted and the diffusion profile can be approximated by a Gaussian function

$$C(y) = C_0\exp\left(-\frac{y^2}{d_{\exp}^2}\right).\tag{5}$$

In this case, $d_{\exp} = \sqrt{4Dt}$ describes the 1/e-depth of the profile. The surface concentration C_0 is below the solubility and given by

$$C_0 = \frac{\tau\,C_{\max}}{\sqrt{\pi Dt}}.\tag{6}$$

The temperature dependence of the diffusion coefficient D is given by the Arrhenius relation

$$D(T) = D_0\exp\left(-\frac{E_A}{k_B T}\right).\tag{7}$$

Here, D_0 is the diffusion constant, E_A the activation energy, k_B the Boltzmann constant, and T is the temperature used for in-diffusion.

As the solubility describes the equilibrium between Yb in LiNbO$_3$ and the pure metallic Yb on the surface, the concentration at the surface can also be written in terms of an Arrhenius-type equation as

$$C_{\max}(T) = \hat{C}\exp\left(-\frac{\Delta H}{k_B T}\right),\tag{8}$$

where ΔH is the mixing enthalpy and $\hat{C}$ is a pre-factor.

Figure 2 shows the high-resolution images of the crystal surfaces obtained using a confocal microscope and applying the differential interference contrast method for the two crystal orientations. When the temperature used for annealing is less than 1130 °C, an increased surface roughness caused by insufficient diffusion is visible. For the x-cut surfaces, the arithmetical mean height (*Sa*) slightly decreases from 12 to 10 nm when the temperature increases from (a) 1000 to (b) 1060 °C.

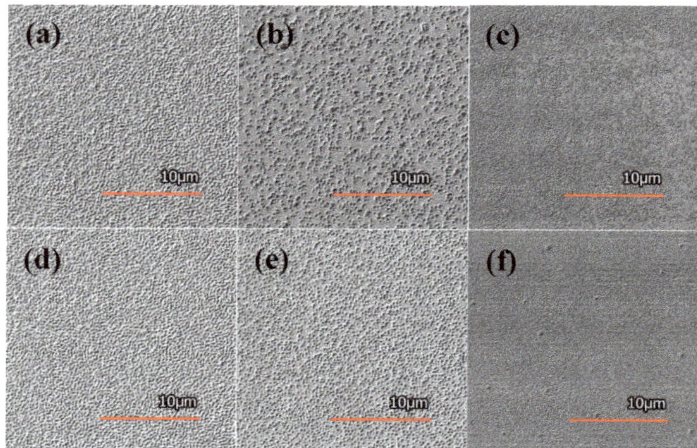

Figure 2. Confocal microscope images of LiNbO$_3$ surfaces coated with 20 nm thick Yb films after annealing for 30 h. Results for x-cut surfaces annealed at (**a**) 1000, (**b**) 1060, and (**c**) 1130 °C; results for z-cut surfaces annealed at (**d**) 1000, (**e**) 1060, and (**f**) 1130 °C.

For the z-cut surface shown in (d) and (e), the *Sa* is reduced from 11 to 7 nm. Annealing for 30 h at 1130 °C results in a smooth surface with *Sa* = 2 nm for both (c) x- and (f) z-cut surfaces, indicating that the Yb ion source at the surface is completely exhausted, and the Yb profile can be described by the thin-film diffusion regime of Equation (5).

For calibration of the SNMS data, the samples annealed at 1130 °C were used assuming a linear relation between ion yield and concentration as well as mass conservation in the diffusion process. For lower temperatures, the experimentally obtained profiles can be best fitted by complementary error functions, as can be seen in Figure 3.

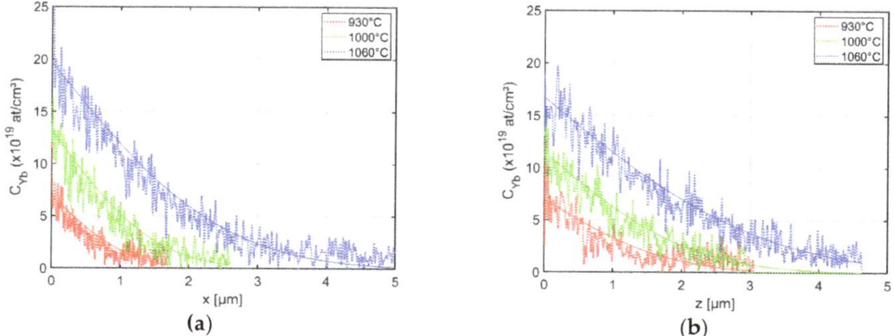

Figure 3. SNMS depth profiles of Yb in LiNbO$_3$ for in-diffused layers with an initial thickness of 20 nm annealed for 30h at different temperatures for (**a**) x-cut and (**b**) z-cut substrates. The dotted lines show the measured profiles and the solid lines indicate complementary error function fits.

The profiles obtained using the highest temperature were fitted by Gaussian functions; see Figure 4. The results obtained for the surface concentration and for the depths d_{erf} or d_{exp} are summarized in Table 1.

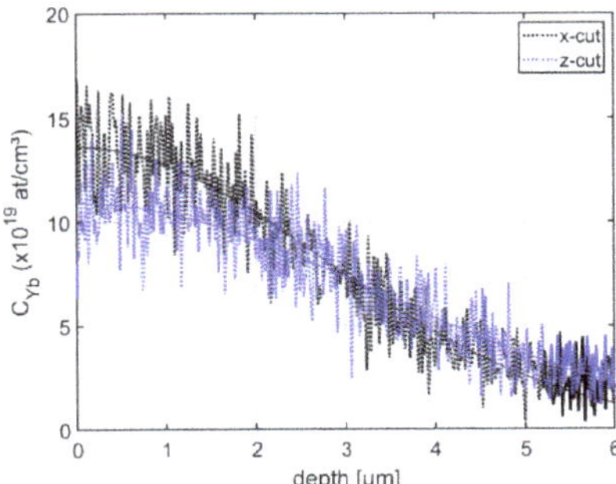

Figure 4. SNMS depth profiles of Yb in LiNbO$_3$ for in-diffused layers with an initial thickness of 20 nm annealed for 30 h at 1130 °C for different cut directions.

Table 1. Parameters obtained for the Yb diffusion in LiNbO$_3$.

Temperature (°C)	Time (h)	$C_0 \times 10^{19}$ (ions/cm^3)	$C_{max} \times 10^{19}$ (ions/cm^3)	d_{erf} (µm)	d_{exp} (µm)
z-cut					
930	30	6.96 ± 0.10		1.99 ± 0.13	
1000	30	11.53 ± 0.09		2.32 ± 0.14	
1060	30	16.89 ± 0.10		3.47 ± 0.19	
1130	30		10.94 ± 0.05		4.68 ± 0.26
x-cut					
930	30	6.89 ± 0.10		1.24 ± 0.08	
1000	30	13.21 ± 0.12		1.66 ± 0.10	
1060	30	20.06 ± 0.12		2.70 ± 0.15	
1130	30		13.62 ± 0.06		3.84 ± 0.21

With the values for the diffusion depth, the diffusion coefficients D are calculated. The temperature dependence of the diffusion coefficient is given by Equation (7). Figure 5 shows the corresponding Arrhenius plots for the two crystal-cut directions used. By correlating the fitted lines with Equation (7), the values for the diffusion coefficients and the activation energy for the x- and z-direction were obtained. The results are summarized in Table 2.

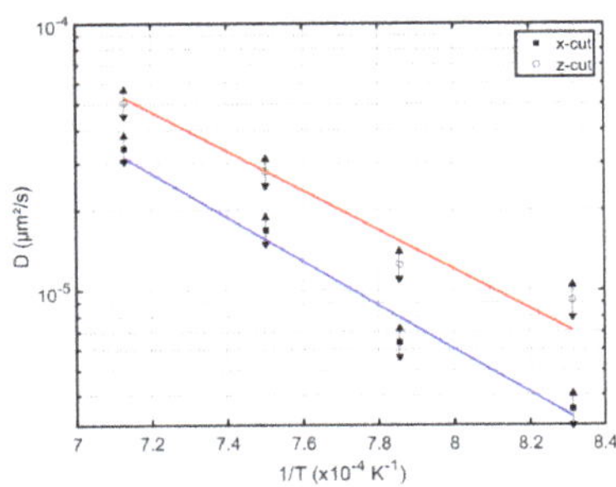

Figure 5. Arrhenius plot of the diffusion coefficients of Yb in LiNbO$_3$ for diffusion parallel to the x- and z-axis. The fitted lines correspond to Equation (7).

Table 2. Diffusion constants and activation energies for Yb diffusion in LiNbO$_3$.

	D_0 ($\times 10^{-8}$ cm^2/s)	E_A (eV)
Diffusion in z-direction	9.7 ± 0.8	1.46 ± 0.07
Diffusion in x-direction	24.1 ± 1.9	1.64 ± 0.07

The temperature dependence of the surface concentration is presented in Figure 6. As can be seen, the measured values for the surface concentration of Yb for the three lowest temperatures indicate that, in contrast to the diffusivity, the solubility of Yb in LiNbO$_3$ shows no anisotropic behavior. The solid line is a fit to the data with Equation (8). This way, the values of $\Delta H = (1.04 \pm 0.23)$ eV and $\hat{C} = (1.6 \pm 0.35) \times 10^{24}$ cm^{-3} were found.

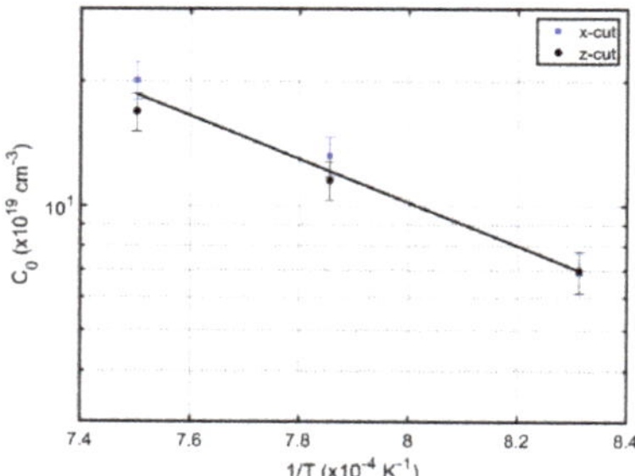

Figure 6. Arrhenius plot of the solubility data of Yb in LiNbO$_3$. The line is a fit of the data using Equation (8).

With these characteristic values for the diffusion of Yb in LiNbO$_3$, a first test was performed for diffusion doping of a ridge waveguide prepared in an x-cut LiNbO$_3$ substrate oriented parallel to the y-axis by diamond blade dicing. Figure 7a shows the simulated Yb concentration profile in the ridge for the case of two 22 nm thick Yb layers being coated under ±60° with respect to the x-axis obtained by numerically solving Fick's second law. For in-diffusion, the sample is annealed for 216 h at 1125 °C. The concentration profile also shows that at the upper right and left corners of the ridge, the expected concentration is below the solubility limit. Thus, no additional surface defects are to be expected. To achieve waveguiding, the upper part of the ridge is additionally doped by in-diffusion of Ti.

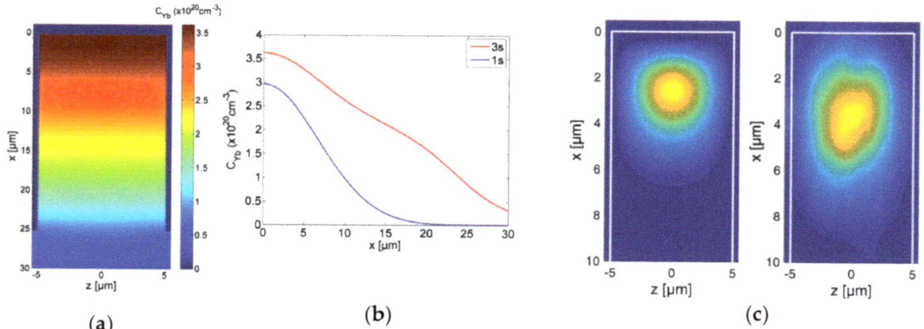

(a)   **(b)**   **(c)**

Figure 7. (a) Simulated Yb profile in LiNbO$_3$ for the in-diffusion of two 22 nm thick layers evaporated under ±60° and in-diffused for 216 h at 1125 °C. (b) The Yb profile in x-direction in the middle of the ridge (red line). The results for a plane layer of Yb diffused in from the surface only are included as a blue line. (c) Simulated (left) and measured (right) intensity distribution for the fundamental mode at 980 nm for a 10 μm-wide ridge-waveguide formed by three-side Ti in-diffusion.

Figure 7c shows the expected intensity distribution of the quasi-TE modes at 800 nm and the intensity distribution measured at 796 nm. It is clearly visible that a good overlap between the Yb profile and the intensity distribution of the mode is achieved. The simulated concentration profiles shown in Figure 7a,b prove that ridge waveguides can be almost homogeneously doped by three-side in-diffusion while avoiding any surface damage that might be caused by exceeding the maximum solubility of Yb in LiNbO$_3$.

The transmission spectrum of such a 10 µm wide and 2.2 cm long ridge waveguide was measured using unpolarized white light coupled into the waveguide by means of a 40× microscope objective. At the output facet, the transmitted light was collected and coupled into a spectrometer. To measure absorption as a function of polarization, a polarizer was placed in front of the collimator of the spectrometer. The measured absorption spectra (π and σ polarization) show three main peaks at 918, 956, and 980 nm (see Figure 8a), which is in good agreement with the absorption spectra reported for bulk-doped Yb:LiNbO$_3$ [22] and Yb:Ti:LiNbO$_3$ channel waveguides [14]. The losses of the waveguide of 0.8 dB/cm were determined at 1064 nm using the Fabry–Perot method. Fluorescence spectra (π and σ polarization) were measured for pumping with a titanium-sapphire laser at 918 nm. Figure 8b shows the polarized fluorescence spectrum. In agreement with previously reported results for Yb:LiNbO$_3$ waveguides [14,20], we observed the expected four main peaks around 960, 980, 1008, and 1062 nm.

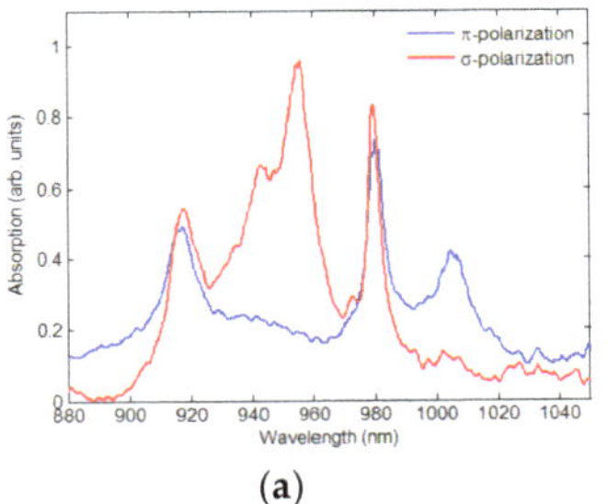
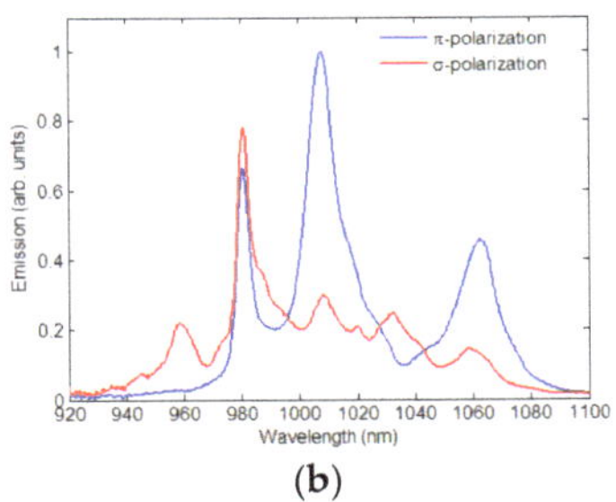

(a) **(b)**

Figure 8. Polarization-dependent absorption (**a**) and fluorescence (**b**) spectrum of a 10 µm wide and 2.2 cm long Yb:Ti:LiNbO$_3$ ridge waveguide.

4. Summary

The diffusion of Yb in LiNbO$_3$ can be described by Fick's law of diffusion. The experimental values determined in this work reveal an anisotropic diffusion coefficient with higher values for the diffusion along the crystallographic c-axis (z-axis). No anisotropy was found for the (maximum) solubility of Yb in LiNbO$_3$. However, the solubility is temperature-dependent. When the diffusion reservoir is not exhausted, small grains of Yb remain on top of the surface, leading to high surface roughness, which has to be strictly avoided when fabricating optical waveguides. Absorption and fluorescence measurements for Yb diffusion-doped ridge waveguides are in good agreement with results published for volume-doped bulk substrates. These results are essential for the future development of efficient Yb-based laser sources in LiNbO$_3$ ridge waveguides.

Author Contributions: Investigation C.E.R., D.B., and S.S.; writing—original draft preparation, C.E.R.; writing—review and editing, D.B., S.S., and D.K.; supervision, D.K.; project administration, D.K.; funding acquisition, D.K. All authors have read and agreed to the published version of the manuscript.

Funding: This research was funded by Deutsche Forschungsgemeinschaft (grant DFG Ki482/17-1).

Conflicts of Interest: The authors declare no conflict of interest.

References

1. Hu, H.; Ricken, R.; Sohler, W.; Wehrspohn, R.B. Lithium Niobate Ridge Waveguides Fabricated by Wet Etching. *IEEE Photon Technol. Lett.* **2007**, *19*, 417–419. [CrossRef]
2. Hu, H.; Ricken, R.; Sohler, W. Low-loss ridge waveguides on lithium niobate fabricated by local diffusion doping with titanium. *Appl. Phys. A* **2010**, *98*, 677–679. [CrossRef]
3. Volk, M.F.; Suntsov, S.; Rüter, C.E.; Kip, D. Low loss ridge waveguides in lithium niobate thin films by optical grade diamond blade dicing. *Opt. Express* **2016**, *24*, 1386–1391. [CrossRef]
4. Courjal, N.; Guichardaz, B.; Ulliac, G.; Rauch, J.-Y.; Sadani, B.; Lu, H.; Bernal, M.-P. High aspect ratio lithium niobate ridge waveguides fabricated by optical grade dicing. *J. Phys. D Appl. Phys.* **2011**, *44*, 305101. [CrossRef]
5. Gerthoffer, A.; Guyot, C.; Qiu, W.; Ndao, A.; Bernal, M.-P.; Courjal, N. Strong reduction of propagation losses in LiNbO$_3$ ridge waveguides. *Opt. Mater.* **2014**, *38*, 37–41. [CrossRef]
6. Cai, L.; Wang, Y.; Hu, H. Low-loss waveguides in a single-crystal lithium niobate thin film. *Opt. Lett.* **2015**, *40*, 3013–3016. [CrossRef] [PubMed]
7. Becker, C.; Oesselke, T.; Pandavenes, J.; Ricken, R.; Rochhausen, K.; Schreiber, G.; Sohler, W.; Suche, H.; Wessel, R.; Balsamo, S.; et al. Advanced Ti:Er:LiNbO$_3$ waveguide lasers. *IEEE J. Sel. Top. Quantum Electron.* **2000**, *6*, 101–113. [CrossRef]
8. Lallier, E.; Pocholle, J.; Papuchon, M.; Grezes-Besset, C.; Pelletier, E.; De Micheli, M.; Li, M.; He, Q.; Ostrowsky, D. Laser oscillation of single-mode channel waveguide in Nd:MgO:LiNbO$_3$. *Electron. Lett.* **1989**, *25*, 1491. [CrossRef]
9. De Micheli, M.; Lallier, E.; Grezes-Besset, C.; Pelletier, E.; Pocholle, J.P.; Li, M.J.; He, Q.; Papuchon, M.; Ostrowsky, D.B. Nd:MgO:LiNbO$_3$ waveguide laser and amplifier. *Opt. Lett.* **1990**, *15*, 682–684. [CrossRef]
10. Di Paolo, R.E.; Cantelar, E.; Pernas, P.; Pedrola, G.L.; Cussó, F. Continuous wave waveguide laser at room temperature in Nd[sup 3+]-doped Zn:LiNbO[sub 3]. *Appl. Phys. Lett.* **2001**, *79*, 4088–4090. [CrossRef]
11. Siebenmorgen, J.; Calmano, T.; Petermann, K.; Huber, G.; Huber, G. Highly efficient Yb:YAG channel waveguide laser written with a femtosecond-laser. *Opt. Express* **2010**, *18*, 16035–16041. [CrossRef] [PubMed]
12. Okhotnikov, O.G.; Gomes, L.; Xiang, N.; Jouhti, T.; Grudinin, A.B. Mode-locked ytterbium fiber laser tunable in the 980-1070-nm spectral range. *Opt. Lett.* **2003**, *28*, 1522–1524. [CrossRef]
13. Malinowski, A.; Piper, A.; Price, J.H.V.; Furusawa, K.; Jeong, Y.; Nilsson, J.; Richardson, D. Ultrashort-pulse Yb3+-fiber-based laser and amplifier system producing >25-W average power. *Opt. Lett.* **2004**, *29*, 2073–2075. [CrossRef] [PubMed]
14. Jones, J.K.; De Sandro, J.P.; Hempstead, M.; Shepherd, D.P.; Large, A.C.; Tropper, A.C.; Wilkinson, J. Channel waveguide laser at 1 μm in Yb-indiffused LiNbO$_3$. *Opt. Lett.* **1995**, *20*, 1477–1479. [CrossRef] [PubMed]
15. Fujimura, M.; Tsuchimoto, H.; Suhara, T. Yb:LiNbO$_3$ Annealed/Proton-Exchanged Waveguide Lasers Pumped by InGaAs Laser Diode at 980 nm Wavelength. *Jpn. J. Appl. Phys.* **2007**, *46*, 5447–5449. [CrossRef]
16. Pal, S.; Das, B.K.; Sohler, W. Photorefractive damage resistance in Ti:PPLN waveguides with ridge geometry. *Appl. Phys. A* **2015**, *120*, 737–749. [CrossRef]
17. Suntsov, S.; Rüter, C.E.; Kip, D. Er:Ti:LiNbO$_3$ ridge waveguide optical amplifiers by optical grade dicing and three-side Er and Ti in-diffusion. *Appl. Phys. B Laser Opt.* **2017**, *123*, 118. [CrossRef]
18. Kip, D.; Brüske, D.; Suntsov, S.; Rüter, C.E. Efficient ridge waveguide amplifiers and lasers in Er-doped lithium niobate by optical grade dicing and three-side Er and Ti in-diffusion. *Opt. Express* **2017**, *25*, 29374.
19. Brüske, D.; Suntsov, S.; Rüter, C.E.; Kip, D. Efficient Nd:Ti:LiNbO$_3$ ridge waveguide lasers emitting around 1085 nm. *Opt. Express* **2019**, *27*, 8884–8889. [CrossRef]
20. Fujimura, M.; Tsuchimoto, H.; Suhara, T. Yb-diffused LiNbO/sub 3/ annealed/proton-exchanged waveguide lasers. *IEEE Photon Technol. Lett.* **2004**, *17*, 130–132. [CrossRef]
21. Amin, J.; Aust, J.; Veasey, D.; Sanford, N. Dual wavelength, 980 nm-pumped, Er/Yb-codoped waveguide laser in Ti:LiNbO$_3$. *Electron. Lett.* **1998**, *34*, 456. [CrossRef]
22. Burns, G.; O'Kane, D.F.; Title, R.S. Optical and Electron-Spin-Resonance Spectra ofYb3+,Nd3+, andCr3+in LiNbO$_3$ and LiTaO$_3$. *Phys. Rev.* **1968**, *167*, 314–319. [CrossRef]

MDPI

St. Alban-Anlage 66

4052 Basel

Switzerland

Tel. +41 61 683 77 34

Fax +41 61 302 89 18

www.mdpi.com

Applied Sciences Editorial Office

E-mail: applsci@mdpi.com

www.mdpi.com/journal/applsci